World Food Supply

This is a volume in the Arno Press collection

World Food Supply

See last pages of this volume for a complete list of titles.

THREE WORLD SURVEYS BY THE FOOD AND AGRICULTURE ORGANIZATION OF THE UNITED NATIONS

ARNO PRESS
A New York Times Company
New York — 1976

Editorial Supervision: MARIE STARECK

Reprint Edition 1976 by Arno Press Inc.

Reprinted from copies in The Princeton University Library

WORLD FOOD SURVEY
ISBN for complete set: 0-405-07766-1
See last pages of this volume for titles.

Manufactured in the United States of America

Library of Congress Cataloging in Publication Data

Food and Agriculture Organization of the United Nations.
Three world surveys.

(World food supply)
Reprint of World food Survey, first published in 1946 by FAO, Washington; of Second world food survey, first published in 1952 by FAO, Rome; and of Third world food survey, first published in 1963 by FAO, Rome.
1. Food supply. 2. Nutrition. I. Title.
II. Series.
HD9000.5.F5825 1976 338.1'9 75-27639
ISBN 0-405-07779-3

CONTENTS

Food and Agriculture Organization of the United Nations
WORLD FOOD SURVEY. Washington, D. C., 1946

Food and Agriculture Organization of the United Nations
SECOND WORLD FOOD SURVEY. Rome, 1952

Food and Agriculture Organization of the United Nations
THIRD WORLD FOOD SURVEY, Freedom From Hunger Campaign Basic Study No. 11, Rome, 1963

Food and Agriculture Organization
of the United Nations

World Food Survey

Washington, 5 July 1946

TABLE OF CONTENTS

1. BACKGROUND OF THE SURVEY

IT IS well known that there is much starvation and malnutrition in the world. Millions of people never get enough to eat, and much larger numbers, not actually hungry, do not obtain the kind of diet necessary for health. Vague knowledge that this situation exists is not enough; facts and figures are needed if the nations are to attempt to do away with famine and malnutrition—an attempt to which they are pledged through the Food and Agriculture Organization of the United Nations.

What is the actual food consumption of the different peoples of the world? How does it compare with their needs? Where are the most serious shortages? What practical goals can be set to remedy these shortages within a reasonable time? What additional quantities of food would be required to reach the goals? These are the points on which more definite information is needed if the ideal of freedom from famine and malnutrition, with all their attendant evils, is to be translated into workable plans.

Exact knowledge on these questions, country by country throughout the world, has been meager except in a few cases. Yet a good deal of material exists in scattered sources, and there are international and national agencies whose business it has been to collect such material. Experts in these agencies have been accustomed to dealing critically with the available information, and they are capable of filling in some of the gaps with estimates based on wide experience.

Early in 1946 several of these agencies loaned the services of some of their staff members to the Food and Agriculture Organization for the purpose of making a world food survey in which the best available figures and estimates would be brought together, critically examined, and reduced to a uniform basis. The objective was to obtain as clear a picture as possible of the world food situation as it was in the years just before the war. FAO needed these figures as a guide in working out proposals for future world food and agricultural policies.

This report gives the results of the survey. It covers 70 countries whose people make up about 90 percent of the earth's population.

It need scarcely be said that the figures for many countries are highly imperfect. Statistical services in most countries will have to be vastly improved before complete and accurate data are obtainable; it is one of FAO's functions to help bring about this improvement, which will take many years.

Some of the authorities who were consulted about the survey, however, have expressed the opinion that the figures brought together here could not have been greatly improved, on the basis of the available information, if the groups of experts had worked for three years instead of the three months or so actually spent on the task. Total calorie estimates, in particular, are probably accurate within 5 percent in most instances for countries with an average intake of 3000 calories a person a day, and within 10 percent for those with an average of around 2000. It should be added that the estimates were submitted for checking to each of the countries concerned, and various adjustments have been made as a result.

In other words, though the figures in the survey must be regarded as provisional and incomplete (their shortcomings are fully discussed in Appendix II on page 32), they give a more accurate and detailed picture of the prewar food situation than has hitherto been available. They are close enough to the truth to be used, with due caution, as a yardstick by which to measure changes that will be required if we are not to return to the unsatisfactory food situation which existed in the years before the war but to have a food and agriculture policy that will meet human needs.

Another question besides those mentioned above must be considered. What would be involved in producing the additional quantities of food needed to meet practical goals? This also is discussed in the report.

The world food survey would not have been possible without the generous assistance of the persons and agencies listed by name in Appendix I, page 31. FAO's debt to them is inadequately expressed in this brief acknowledgment.

2. THE PREWAR FOOD PICTURE

FOOD supplies available in 70 countries in the prewar period have been estimated. The results are summarized in figure 1, opposite page 24 in terms of calories per head of population daily. The calorie figures are presented in the form of a bar chart which shows the food consumption of the different countries at a glance. Each long bar on the left represents the total number of calories per caput daily furnished by the food supplies (production plus imports minus exports) available in each country. The long bars are divided to show the number of calories yielded by nine different groups of foods: cereals; roots and tubers; sugar; fats; pulses (peas and beans); fruits and vegetables; meat, fish, and eggs; milk; and wine and beer.

Estimates of supplies in terms of calories are convenient for purposes of comparison. It should be borne in mind that the more concentrated foods, such as cereals, fats, sugar, and meat have a higher calorie value per gram or ounce than less concentrated foods such as fruits and vegetables, which usually contain a good deal of water; many grams of a leafy vegetable are required to furnish the same number of calories as one gram of rice.

The short bars on the right in figure 1 show the amount of protein per caput daily (in grams, not calories) supplied by the national diets. These bars are divided to show the amount of protein obtained from foods of animal origin and the amount obtained from vegetable foods such as grains and pulses. A diet rich in animal protein, especially in protein derived from milk and milk products, is likely to contain enough of the foods rich in minerals and vitamins to supply these nutrients in adequate amounts.

The great variation in the dietary patterns of the different countries is obvious.

Table 1, Appendix III gives per caput figures for total calories and calories from the different food groups. Table 2 shows food supplies in terms of kilograms per head per week. Both chart and tables represent food supplies at the retail stage or its equivalent. This means that allowances have been made for losses and wastage at various stages between production on the farm and entry into the retail stage, but none from this point onwards.

The Extent of Malnutrition

Calculations based on the prewar populations of the individual countries show that in the years before the war—

In areas containing over half the world's population, food supplies at the retail level (not actual intake) were sufficient to furnish an average of less than 2250 calories per caput daily.

Food supplies furnishing an average of more than 2750 calories per caput daily were available in areas containing somewhat less than a third of the world's population.

The remaining areas, containing about one-sixth of the world's population, had food supplies that were between these high and low levels.

The proportion of the world's population in each of the three groups is shown in figure 2.

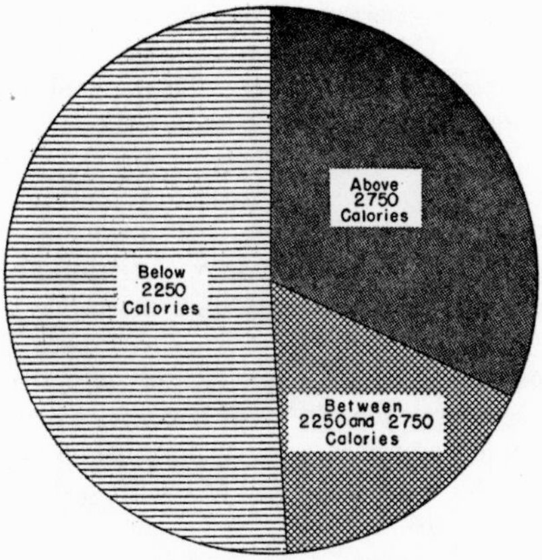

FIG. 2 RELATIVE SIZE OF WORLD POPULATIONS AT DIFFERENT CALORIE LEVELS

The high-calorie areas include most of the western world, all of North America, and much of Europe. Oceania and the Union of Soviet Socialist Republics also belong to this group, but it includes only three countries in South America.

The medium-calorie areas include most of southern Europe, three countries in Asia, a part of the Middle East, a part of Africa, and a part of South America.

The low-calorie areas include most of Asia, a part of the Middle East, all of Central America, and probably parts of South America and of Africa which were not covered by this survey.

But averages do not tell the whole story. They conceal many sharp differences. Whatever the average calorie level of a country, some people obtain considerably more than the average while a large number have less. Even in the countries with the most liberal food supplies and the highest calorie intake, it is well known that a considerable part of the population is not well nourished. There are wide local variations in food supply which are not brought out by average figures for the whole population.

The Areas of Greatest Deficiency

The areas of greatest deficiency, according to this survey, are Central America and most of Asia. They also probably include parts of South America and of Africa not covered by the survey.

Many of the low-calorie countries are in the tropics and subtropics. In these countries, food energy requirements may be less than in colder countries. The average size of the people is usually smaller. The proportion of children in the population also is usually greater because of high birth and death rates. These factors, however, cannot account for the great difference in per caput daily calorie intake between the low and the high countries. A population with a high percentage of children, for example, would require only some 100–150 calories a person a day less (but relatively more minerals and vitamins) than a population with the age composition now typical of western civilization. As has been seen, the actual difference is around 1000 calories a person a day. Calorie intake in the low-calorie countries is only two-thirds of that in the high-calorie countries.

A vicious cycle probably influences the nutrition of peoples who subsist at a low level of calorie intake. Restricted diet leads to poor physique and a lowering of basal metabolism and energy output, and these in turn facilitate adaptation to a restricted diet. The cycle would be reversed if the food supplies of these people were increased and improved. Their physical development would be better and their energy output raised, and hence they would require more food.

It is evident from this survey that about half of the world's population was subsisting before the war at a level of food consumption which was not high enough to maintain normal health, allow for normal growth of children, or furnish enough energy for normal work. Poor nutrition is associated with high death rates and a low expectation of life, high mortality in infancy and early childhood and among women during the child-bearing period, increased susceptibility to many diseases such as tuberculosis, and impaired working capacity. It is not necessary to enlarge further on this point, which has been discussed many times, for example in publications of the League of Nations and in the reports of the Hot Springs Conference and of the United Nations Interim Commission on Food and Agriculture.

The Composition of the Food Supply by Food Groups

Estimates of calories per caput obtained from the different food groups are subject to a wider range of error than estimates of total calories. Many gaps in the statistical data need to be filled, and intake figures do not show the great variations in local food supplies and food habits which only painstaking study in the future can reveal. Nevertheless, the tables and the chart clearly demonstrate the marked differences in the quality of the national diets of the countries included in this report.

Countries and areas with average calorie levels around the 3000 mark or more had well balanced national diets. These included North America, Oceania, Argentina, the British Isles, Scandinavia, the Netherlands, Switzerland, and Germany. In all these the consumption of cereals in relation to that of other foods was relatively low—about 1000 calories daily were obtained from cereals—and milk and meat consumption comparatively high, giving a supply of animal protein of about 50 grams. In Iceland, where environment conditions the food supply as in all arctic or subarctic areas, intake of animal protein (63 grams) was higher than in any other country surveyed except Argentina and Uruguay, mainly because of the high intake of fish—the catch is equivalent to two tons a head a year. Argentina, Uruguay, and Paraguay, the meat-producing countries of South America, together with Australia and New Zealand, were the only other countries with substantially more than 50 grams of animal protein per caput daily.

This general dietary pattern contrasts sharply with that in countries where the average total calorie supplies were around the 2000 mark or less. These countries were: in the Far East—India, Java, the Philippines, Korea; in the Middle East—Iran, Iraq, and Transjordan; in Central America—Mexico, El Salvador, and Costa Rica; and in South America, Colombia. Here the low average energy value of the diet reflects widespread poverty. A high proportion of total calories was obtained from cheap foods rich in carbohydrates, such as cereals. With very few exceptions the average calorie value from the cereal

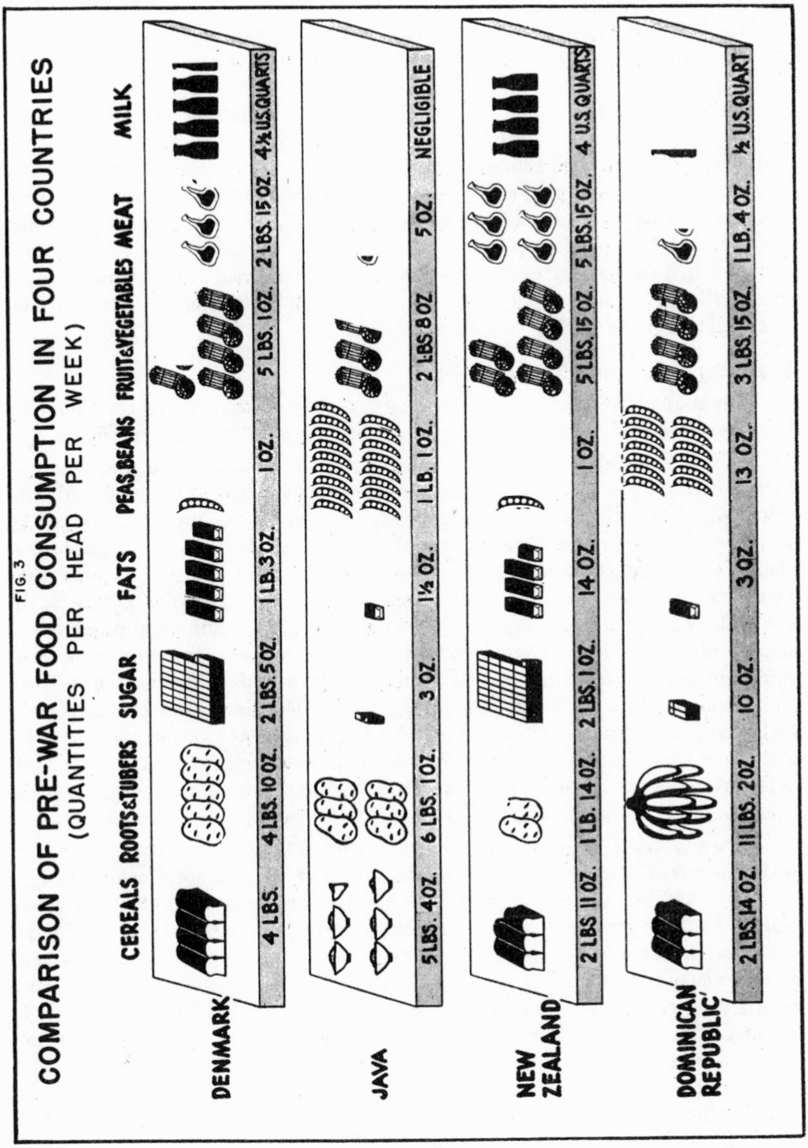
FIG. 3
COMPARISON OF PRE-WAR FOOD CONSUMPTION IN FOUR COUNTRIES
(QUANTITIES PER HEAD PER WEEK)
CEREALS
ROOTS & TUBERS
SUGAR
FATS
PEAS, BEANS
FRUIT & VEGETABLES
MEAT
MILK
DENMARK
4 LBS.
4 LBS. 10 OZ.
2 LBS. 5 OZ.
1 LB. 3 OZ.
1 OZ.
5 LBS. 1 OZ.
2 LBS. 15 OZ.
4½ U.S. QUARTS
JAVA
5 LBS. 4 OZ.
6 LBS. 1 OZ.
3 OZ.
1½ OZ.
1 LB. 1 OZ.
2 LBS. 8 OZ.
5 OZ.
NEGLIGIBLE
NEW ZEALAND
2 LBS. 11 OZ.
1 LB. 14 OZ.
2 LBS. 1 OZ.
14 OZ.
1 OZ.
5 LBS. 15 OZ.
5 LBS. 15 OZ.
4 U.S. QUARTS
DOMINICAN REPUBLIC
2 LBS. 14 OZ.
11 LBS. 2 OZ.
10 OZ.
3 OZ.
13 OZ.
3 LBS. 15 OZ.
1 LB. 4 OZ.
½ U.S. QUART

supply was never substantially below 1000. Two exceptions are Costa Rica, which as a sugar-producing country obtained a large part of its total carbohydrates from this source, and Colombia, which, though it grew maize and had to import its wheat and rice, obtained much of its carbohydrates from potatoes, bananas, and plantains.

It is interesting that in the two extreme cases considered, with average total calories of 3000 and 2000 respectively, the calories obtained from cereals approximated 1000.

Of the countries in which the average total calorie intake was intermediate, those with a low cereal consumption, such as Paraguay with 545 cereal calories and Kenya-Uganda with 586, consumed large quantities of such roots and tubers as sweetpotatoes and plantains. In this intermediate group countries with exceptionally high cereal intake form a geographically continuous area stretching from eastern Europe through the Union of Soviet Socialist Republics to Manchuria. In these countries the quantity of food obtained from animal sources was small and cereals supplied as much as 50 to 60 percent of the total calories, to the detriment of variety and balance.

Examples of Different National Diets

Four countries, New Zealand, Denmark, Java, and the Dominican Republic, have been selected to illustrate differences in national food supplies (figure 3).

In New Zealand, with a high average food consumption, the diet was well balanced. Calories from cereals amounted to a little less than 1000 and consumption of meat, milk, and fat was high. The supply of protein averaged 96 grams, of which 65 percent was of animal origin. Denmark was the highest food consumer among the Scandinavian countries. It is of interest to note that Denmark and New Zealand, though situated on opposite sides of the earth and differing in many characteristics of national life, consumed approximately similar kinds of diets. Cereal consumption was equally low and milk consumption equally high in the two countries. The main differences were that while consumption of meat, fish, and eggs in Denmark was comparatively high, the consumption of these foods in New Zealand was twice as great; but on the other hand Denmark consumed half again as much fat as New Zealand. When there is abundance and variety of food and purchasing power is high, countries tend to choose a diet fully adequate for health.

Java and the Dominican Republic, by contrast, are examples of countries with low average levels of consumption. In Java, with a total calorie supply of about 2000, the calories furnished by cereals were more than 1000 per caput daily. Carbohydrate intake was further increased by the consumption of large quantities of cassava, so that not only was the average supply of animal protein almost negligible (4 grams), but the total protein (43 grams) was the lowest recorded in all the 70 countries surveyed. The Dominican Republic was little better off; the main difference lay in the larger intake of animal protein, accounted for by the considerably greater consumption of milk, meat, fish, and eggs. Bananas have been included in the "roots and tubers" group in the case of this and other tropical countries in which they were a staple article of diet. In nutritive value they are akin to this group.

Income and Nutrition

Poverty is the chief cause of malnutrition. In the course of preparing this survey prewar calorie consumption was compared with national incomes per person as far as these are known. It is interesting to observe that all the countries in which the supply of calories per caput was less than 2250 a day were countries in which the average per caput income was less than U.S. $100 a year. At the other end of the scale there were 365 million people living in countries in which the average supply of calories exceeded 2900 a person a day. Of these, 342 million were in countries in which the average income exceeded U.S. $200 a person a year.

Thus for the world as a whole it can be said, "Tell me what you earn and I will tell you how you eat." There are exceptions, but in general well-to-do countries fare well nutritionally, poor countries fare badly, and the poorest groups within these countries fare the worst.

3. NUTRITIONAL TARGETS

THE NEXT step after determining world food consumption country by country was to set up nutritional targets showing the changes in food supplies which are necessary to provide the world with a better diet.

The Hot Springs Conference recommended that governments should—

> adopt as the ultimate goal of their food and nutrition policy, dietary standards or allowances based upon scientific assessment of the amount and quality of food, in terms of nutrients, which promote health, and distinguish clearly between these standards and the more immediate consumption goals which necessarily must be based upon the practical possibilities of improving the food supply of their populations.

The problem of dietary standards or allowances for international application is one with which both FAO and the World Health Organization will be concerned in the future. In the United States of America the recommended daily allowances of the National Research Council (NRC) have been widely accepted and are used by official agencies. All such standards are provisional in the sense that they are subject to modification as nutritional knowledge advances. They represent the consensus of experts at a given time.

Where food consumption is already adequate in the quantitative sense, "optimum" standards can be applied to determine the changes in food supply necessary to improve the quality of national diets. But in many of the countries with a medium calorie intake, and in all of those with a low intake, consumption goals must be set considerably below the optimum if they are to be reached within any reasonable time. The achievement of such intermediate goals would bring a vast improvement in world nutrition. They may be regarded as milestones on the road to the ultimate Hot Springs objective.

Early in 1946 FAO convened a small group of nutrition experts (see Appendix I, page 31) to consider the question of targets, with the estimates of prewar food supplies as the starting point. It was agreed that in drawing up targets weight must be given to the present position as regards the production and supplies of the various foods, and that the targets should call for modification in existing dietary patterns rather than revolutionary changes. The group of experts suggested the following principles and methods which could be applied, with the exercise of careful judgment, in approaching the problem:

(a) A per caput calorie intake of 2550–2650 should be taken as the minimum level to which intake should be raised in the low-calorie countries, and the quantities of additional foods required should be estimated on this basis.

(b) *Cereals.* If calories from cereals fall between 1200 and 1800, no change should generally be recommended. If they fall below 1200, and if total calorie intake is below 2600, some increase in cereal intake may be recommended unless the total calories from cereals, starchy roots and tubers and starchy fruits, sugar, fats, and pulses exceed 2000–2100.

If cereal calories exceed 1800 and total calories are high, the question of decreasing the former should be considered. In deciding the quantities of other foods which in such circumstance can replace cereals, weight must be given to the dietary pattern as a whole and the objective of improving nutrition must be kept in view.

(c) *Starchy roots and tubers and starchy fruits* (for example, bananas, which in composition resemble roots and tubers). An intake of 100 to 200 calories from these foods may be taken as a desirable objective. A larger consumption may, however, be advocated if intake of cereals is low and adequate amounts of protein can be obtained from such foods as pulses, milk, meat, and fish. But where these cannot easily be made available, as for example in certain manioc-eating countries, too high a consumption of starchy roots may seriously lower protein intake.

(d) *Sugars*. In general, no increase in the intake of sugars should be recommended. If calories from sugar exceed 10 to 15 percent of total calories, some reduction may be considered, with due regard to the dietary pattern as a whole.

(e) *Fats*. Total daily calories from fats (as a separate food group) should be at least 100 and preferably 150–200. Intake of fat through the medium of other food groups must be taken into consideration.

(f) *Pulses*. In countries in which pulses are already an important feature of the dietary pattern, calories from this source may well reach 250–300 daily. In general, this means countries in which meat supplies are of necessity low (say below 150 calories from meat daily) and sources of animal protein limited. But even when meat calories are as high as 200–250, calories from pulses may be pushed to 200–250 if this is in general conformity with dietary habits. Pulse intake must be considered in relation to intake of cereals, starchy roots and tubers and starchy fruits, milk, and meat.

(g) *Fruits and vegetables*. Total calories from these foods (excluding starchy vegetables and fruits) should be at least 100 per caput daily. Preference must be given to leafy green and yellow vegetables and fruits and to fruits and vegetables which are good sources of vitamin C. The quantities of fruits and vegetables recommended should be considered in relation to their nutritive value. If the kinds grown are of low vitamin content, daily calories from this source should be raised. If the reverse, they can perhaps be slightly reduced.

(h) *Meat (including poultry), fish, and eggs*. Not less than 100 calories per caput daily, and preferably 150–200, should be derived from these sources. If intake of milk and pulses is high, that of this group can be correspondingly reduced. Fish can replace meat in countries in which the latter cannot easily be produced in quantity and where fish supplies can be readily increased.

(i) *Milk and milk products*. An intake of 300–400 calories per caput daily represents a desirable minimum level of consumption. In recommending milk supply targets, weight must, however, be given (1) to existing dietary habits in respect of milk consumption, (2) to the present level of milk intake, and (3) to the possibility of providing certain important nutrients of milk through a combination of pulses and leafy green and yellow vegetables. Small fish eaten whole can supply calcium to replace milk calcium, but this is not the case when only fish muscle is eaten. In countries in which milk

FIG. 4

PRE-WAR FOOD SUPPLIES AND NUTRITIONAL TARGETS IN EIGHTEEN AREAS

(CALORIES PER HEAD DAILY AT THE RETAIL LEVEL)

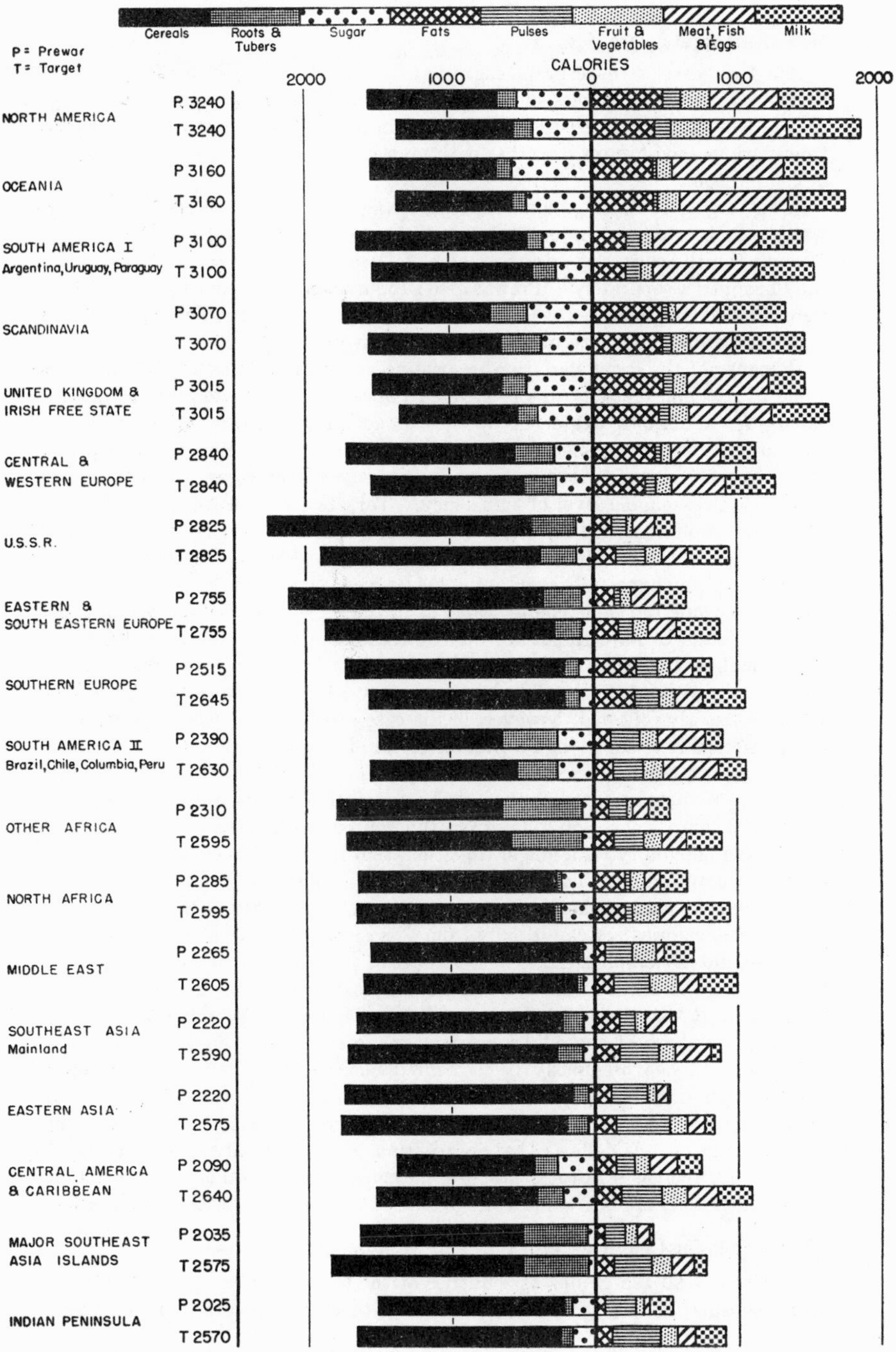

supplies are at present negligible or nonexistent, the milk calorie target may temporarily be set at 50–100 calories, which will represent a very large percentage increase over existing supplies.

How the Targets were Set Up

On the basis of the criteria suggested by the Nutrition Committee, which conform with the principles of modern nutritional science, targets were drawn up for 18 areas. These are given in figure 4, in terms of total calories and calories from various food groups, together with prewar consumption levels for purposes of comparison.

The targets show approximately the changes and increases in food supply which may be aimed at in improving nutrition in 18 areas covering 70 countries. Grouping of the countries into these areas followed the drawing up of targets for the 70 countries separately. Prewar food supplies per caput, country by country, were studied and changes were introduced which would significantly improve national diets without altering their general pattern too widely.

Because of the great variation in existing consumption and in the nutritive value of national diets, it was considered impracticable to put forward targets calling for a uniform degree of nutritional adequacy. In countries in which food supplies were insufficient in quantity as well as unsatisfactory in quality, the first step was to consider the increases in food supplies necessary to raise calorie intake to a reasonable level of sufficiency. Targets for countries in which prewar food supplies yielded less than 2600 calories per caput daily have been adjusted to bring the calorie level to 2600 (plus or minus 50). For countries with calorie supplies above this level, adjustments were made to improve the quality of the diet while the same energy value as in the prewar period was maintained.

The system of adjustment for quality recommended by the nutrition committee makes it possible to suggest desirable changes and improvements in the food supply without drastically altering dietary habits. The recommendations were not exactly followed, however, in the case of areas in which it appeared to be impracticable to set as high a goal for certain food groups as that put forward by the committee. For example, to supply the additional quantities of milk and milk products required in the Far East to attain the desirable minimum of 300 to 400 calories a person a day from this source will be outside the bounds of practicable possibility for a longer time than that likely to be required to achieve other consumption goals called for by the targets, since the prewar consumption of milk in that part of the world was almost negligible. Supplies of milk equivalent to 50 calories per caput daily from this source have been suggested.

It should be emphasized that to improve nutrition in any given area or country more is needed than an increase in total food supplies along the lines suggested. Satisfactory distribution of food within countries so that all obtain an adequate share of the additional food made available is of the utmost importance. As has previously been pointed out, average figures of per caput consumption may conceal wide variations in the consumption of population groups in any country—for example, regional groups or groups of different economic status. It follows that in addition to bringing about the necessary changes and increases in total supplies, measures must be taken to ensure that those whose diets are below the average benefit in accordance with their needs.

The Targets for Countries with Low and Medium Calorie Intake

For nearly all the groups of countries at the low calorie level, an increase in cereal consumption is advocated, principally to raise the total calorie intake.

Substantial increases in the consumption of fats, pulses, fruits and vegetables, milk, and foods in the meat, fish, and eggs group are called for except in a few countries in these groups where prewar consumption of certain of these foods was fairly large—for example, fats in North Africa and meat in Brazil, Chile, and Colombia. For sugar the target is the same as the prewar level of consumption except in the case of the Central American and Caribbean sugar-producing countries where prewar intake was exceptionally high.

The Targets for Medium- and High-Calorie Countries

The targets for eastern and southern Europe and for the Union of Soviet Socialist Republics also call for increases in the consumption of fruits and vegetables and of milk (and meat in the case of the Soviet Republics), but suggest some reduction in cereals. Where cereal consumption is high, it normally declines as the dietary improves.

The targets for North America, the British Isles, Scandinavia, central and western Europe, and Oceania represent adjustments within high-calorie diets rather than an increase in total consumption. The target figures correspond closely with those for a "moderate cost diet" drawn up by nutrition experts in the United States Department of Agriculture, which supplies nutrients in accordance with the recommended allowances of the National Research Council. The target figures also conform in general with the changes in food consumption which in western countries tend to follow a rise in the purchasing power of the lower income groups that permits these groups a freer choice of foods.

It will be seen that reductions in average sugar supply per caput have been suggested by the targets for certain areas in which calories from this source exceeded 10–15 percent of total calories, which represents the quantity of sugar estimated to be sufficient for cooking and adding to food for sweetening. The reason is that when sugar is consumed in large quantities, it often supplants other foods of greater nutritive value. Once a luxury, sugar is now consumed in many countries in four or five times the amounts that were consumed 100 years ago. It should be pointed out that, with increase in population, a reduction in the per caput supplies of any food does not necessarily imply a reduction in the absolute quantities required. The targets, however, indicate that the scope for further increase in sugar consumption is limited, in marked contrast with the fact that desirable increases in milk consumption are so large that the quantities needed will with difficulty be produced after many years of agricultural development.

The targets for the three South American countries with a high calorie intake—Argentina, Paraguay, and Uruguay—provide for increases in the consumption of fruits and vegetables and of milk, and a decrease in that of sugar. No increase in meat consumption is suggested, since this is already high.

The Targets Require Greatly Increased Production

What would the attainment of these targets require in terms of actual quantities of the different foods? The targets indicate the direction to be followed in the improvement of diets, but they do not represent objectives to be reached within any specific period of time. The time requirement, which will depend on the whole economy of the countries concerned and the degree of international cooperation in increasing agricultural production and trade and in raising standards of living, will have to be determined at a later stage when governments consider the measures they are going to take to achieve satisfactory food consumption goals.

TABLE 3

PREWAR FOOD SUPPLIES (P) AND PERCENTAGE CHANGES REQUIRED TO MEET NUTRITION TARGETS BY 1950 (T)—UNITED STATES OF AMERICA AND UNITED KINGDOM

(Amounts in 1,000 metric tons—Changes as percent of prewar supply)

Category		*United States (12% increase in population by 1950)*		*United Kingdom (6% increase in population by 1950)*	
		Amount	*Change*	*Amount*	*Change*
Grain products[1]	P	11,700		4,510	
	T	12,170	+4.0	4,400	−2.5
Roots and tubers[2]	P	8,515		3,700	
	T	9,260	+8.7	4,000	+6.0
Sugar[3]	P	6,280		2,350	
	T	5,490	−12.6	2,045	−13.0
Fats and oils[4]	P	2,635		970	
	T	2,645	+0.4	970	0.0
Pulses, nuts, and cocoa	P	1,140		310	
	T	1,215	+6.6	295	−5.0
Fruits and vegetables[2]	P	26,340		5,330	
	T	39,140	+48.6	9,090	+70.5
Meat, fish, and eggs[5]	P	11,440		4,000	
	T	13,450	+17.6	4,260	+6.5
Milk[6]	P	3,245		820	
	T	5,050	+55.6	1,290	+57.5

SOURCE: Prewar food supplies computed from data given in *Food Consumption Levels in the United States Canada, and the United Kingdom,* Third Report of a Special Joint Committee set up by the Combined Food Board Washington: Production and Marketing Administration, U. S. Department of Agriculture, 1946, table 7, p. 27.

[1]Flour or product basis.

[2]Fresh equivalent.

[3]Sugar content of sugars and sirups.

[4]Fat content. Includes butter.

[5]Including fresh, cured and canned meats (carcass weight), and edible offal; poultry, game, and fish (edible weight); eggs and egg products (fresh equivalent).

[6]Milk and milk products, excluding butter (total milk solids, fat and nonfat).

In considering the quantities of food required by the targets it was necessary, however, to introduce the time element by setting hypothetical dates for their attainment, since increases or decreases in population had to be taken into account as well as changes in consumption per person.

Table 3 shows the additional amounts of food, in relation to prewar supplies, for two high-calorie countries, the United Kingdom and the United States of America, which were already fairly near the targets, with 1950 as the assumed date for reaching them.

Table 4 shows the food supply requirements for a number of countries that were at low or medium calorie levels before the war, with 1960 as the assumed date for reaching the targets. The areas included are China, India, eastern Europe with the exception of Poland (which was omitted because boundary changes make estimates of future population difficult), and three South American countries—Brazil, Chile, and Colombia. It will be noted that, for almost every food group, considerable increases in supplies would be needed over and above those required to meet increases in population. For China and India, with vast populations and low consumption levels, the increases called for in the case of some food groups are enormous.

TABLE 4

PREWAR FOOD SUPPLIES (P) AND PERCENTAGE CHANGES REQUIRED TO MEET NUTRITION TARGETS BY 1960 (T)—FOR FOUR AREAS

(Amounts in 1,000 metric tons—Changes as percent of prewar supply)

Category		*China—22 Provinces (15% increase in population by 1960)*		*India (25% increase in population by 1960)*	
		Amount	*Change*	*Amount*	*Change*
Cereals (whole grain)	P	89,760		64,800	
	T	103,220	+15	90,000	+39
Roots and tubers, fresh	P	17,700		6,780	
	T	29,380	+66	13,760	+103
Sugar	P	610		5,540	
	T	700	+15	6,925	+25
Fats and oils	P	2,290		1,070	
	T	3,620	+58	2,280	+113
Pulses and nuts	P	11,300		8,550	
	T	17,970	+59	15,730	+84
Fruits and vegetables	P	18,300		13,540	
	T	78,140	+327	58,220	+330
Meat, fish, and eggs[1]	P	6,015		3,000	
	T	8,720	+45	12,150	+305
Milk, fluid equivalent	P	20		23,500	
	T	1,150	+5,650	37,600	+60

Category		*Southeastern Europe[2] (10.4% increase in population by 1960)*		*South America[3] (48.6% increase in population by 1960)*	
		Amount	*Change*	*Amount*	*Change*
Cereals (whole grain)	P	11,430		7,110	
	T	11,100	—3	12,090	+70
Roots and tubers, fresh	P	2,215		10,720[4]	
	T	2,790	+26	13,080[4]	+22
Sugar	P	280		1,220	
	T	310	+10	1,815	+49
Fats and oils[5]	P	270		330	
	T	350	+31	545	+65
Pulses, nuts, and cocoa	P	315		1,115	
	T	610	+93	1,900	+70
Fruits and vegetables	P	5,000		9,770	
	T	8,900	+78	16,940	+73
Meat, fish, and eggs[1]	P	1,370		3,130	
	T	1,520	+11	4,975	+59
Milk, fluid equivalent	P	6,020		3,980	
	T	10,650	+77	11,300	+184

[1] Meat (carcass weight), fish (edible portion), eggs (fresh equivalent).
[2] Includes Bulgaria, Hungary, Rumania, Yugoslavia. Adjusted for changes in boundaries and population.
[3] Includes Brazil, Chile, Colombia, and Peru.
[4] Includes bananas as purchased.
[5] Includes butter. Fat content.

TABLE 5

WORLD FOOD NEEDS IN 1960

(Approximate percent increase over prewar supplies required to meet targets, assuming a 25 percent increase in world population)

Commodity	*Percent*
Cereals	21
Roots and tubers	27
Sugar	12
Fats	34
Pulses	80
Fruits and vegetables	163
Meat	46
Milk	100

Finally, table 5 shows, in percentages, the order of increases in supplies required for all the 70 countries in this survey, assuming that the targets were reached by 1960 and that world population has risen 25 percent by that date. This estimate of food needs in 1960 gives some idea of the magnitude of the task to be undertaken and the opportunities ahead for food producers if the nations set out to improve nutrition on a world scale.

The highest increases shown in table 5 are 163 percent for fruits and vegetables and 100 percent for milk and milk products. This indicates the nutritional importance attached to these two food groups. Certain fruits and vegetables are rich in some of the essential vitamins. Milk is a foodstuff which contains all the constituents essential for human life. Further, most of them are present in milk in physiologically balanced proportions, which favors more complete assimilation. Had it been practicable to raise the milk consumption targets of all countries to the desirable level, the aggregate percentage increase would have been considerably higher still. Milk and vegetables are highly perishable unless subjected to fairly complex processing. To a large extent, therefore, they will have to be produced in the areas in which they are consumed.

The nutritive value of pulses, which can be grown comparatively easily, is particularly significant when meat and milk consumption is low. Table 5 indicates an increase of 80 percent in supplies. There are many places where the consumption of meat needs to be increased but where production cannot easily be raised. It would be necessary to make surveys and investigations exploring all the possibilities in order to provide the increase of 46 percent indicated in table 5. In many countries it will be easier to raise intake of animal protein by increasing supplies of fish than by increasing those of meat. With increased meat production there will be a need for more cereals, particularly the coarse grains, for feedstuffs. The 21 percent increase in cereals shown in the table takes account of direct human consumption only.

Targets drawn up after careful investigation of local conditions and resources would unquestionably differ in many respects from those put forward here. Nevertheless, neither the inaccuracies of the food supply data nor the provisional and illustrative nature of the targets affects the broad issues considered in this report. Moreover, the measures needed for raising levels of nutrition throughout the world will be essentially the same whether the period needed for the realization of these or any other series of nutritional targets is as short as 10 or as long as 50 or even more years.

4. THE DIRECTION OF FUTURE ADVANCES

COUNTRIES with inadequate food supplies must obtain the additional food needed to raise nutrition levels by importing from other countries, by producing more themselves, or by a combination of both. Great Britain is an example of the first method. The supplies of food produced by its own agriculture have long been entirely inadequate to meet the requirements of its people, but they have been able to get what they need by exchanging goods and services for food from other countries. During the war, the United Kingdom also increased its own food production by some 70 percent in terms of calories.

Although international trade in food will be increasingly important, the greater part of the additional supplies required by the low-calorie countries to reach the consumption goals suggested in this report will in most cases have to be obtained by expanding their own food production. To understand the extent of the effort needed to achieve this expansion, it is necessary to consider the targets from another angle.

The food supplies of most of the less developed countries consist largely of cereals and other foods of plant origin. The people live mainly on a vegetarian diet (which, however, often lacks sufficient quantities of the nutritionally important vegetables and fruits). In general, the targets require diets containing more foods of animal origin.

When crops are fed to animals instead of being eaten directly by human beings, they lose 80 to 90 percent of their calorie value before they reemerge in the form of meat and milk. For convenience, suppose we adopt the term *original calories*, which is sometimes used for the calories yielded by crops. About seven[1] of these original calories are required to produce one calorie from animal products.

The prewar North American diet contained about 2200 calories per caput daily from foods of plant origin and about 870 calories from livestock products. If the latter figure is multiplied by seven, the total value of the diet in original calories becomes 6090 + 2200, or 8290. At the other end of the scale, the diet of certain islands in Southeast Asia contained about 1940 calories from plant products and only 100 from livestock products, which gives 1940 + 700, or 2640, as the total value in original calories. Thus the value of the North American diet in terms of original calories was about three times that of the diet in Southeast Asia.

The world needs more food both to feed more people and to feed people better. Figure 5 illustrates the relation between the two. It shows in percentages the increases in original calories required by the targets in 1960 and 1970 in a number of the less developed countries, including India, Southeast Asia, and certain other areas, and also the relative demands made by changes in the quality of diets and by the growth of population if the latter continues to expand at about the present rate, which for most of the less developed countries is 10 to 20 percent a decade.

By 1960 original calories would have to be increased by 90 percent in comparison with the prewar value. Fifty-five percent of this increase is accounted for by improvements in the diet and 35 percent by population growth. By 1970, the increase in original calories required would be about 110 percent.

There is a rough relation (see figure 6) between the value of a diet in original

[1]This figure applies to central and western Europe. The multiplier varies, of course, from one class of livestock to another and also among countries, according to differences in the quality of livestock and in feeding practices. Further research is needed on this subject, but the argument is essentially the same whether the multiplier is 4, 7, 10, or some other number.

calories and the agricultural resources needed to produce it, using the term *resources* in a broad sense to include everything concerned in the production of food from land.

Clearly, to double the food supplies, in terms of original calories, in the less developed countries will require a great expansion in agricultural resources, and indeed in all other resources as well. Large increases in imports may be needed also. That would call for expanded production in exporting countries as well as production of commodities in the importing countries to trade for food. Nothing less is involved than a transformation of life in all its aspects which challenges the best efforts of science and industry, governments and peoples.

Although these broad issues have been discussed in reports of the Hot Springs Conference, the Interim Commission, and FAO, it is worth going over some of the main points briefly in the present context.

Improvements in Farming Efficiency

Probably the first need, if there can be said to be any priority in the needed changes, is to improve farming efficiency so as to increase the yields per unit of land. This means in particular making greater use of fertilizers, growing better varieties of crop plants, controlling pests and parasites, and having efficient tools and machinery. The high yields obtained on experimental farms in the less developed countries show how much can be done by these means.

In the case of India it has been estimated that per acre yields of grain could be increased by 30 percent in ten years—5 percent by the use of improved varieties, 20 percent by manuring, and 5 percent by protection against pests.[1] Subsequent improvements, it is estimated, would probably bring the total increase to 50 percent.

A committee of soil and fertilizer experts convened by FAO in connection with the world food survey estimated that India could advantageously use 1.5 million tons of nitrogen, 750,000 tons of phosphate, and 150,000 tons of potash annually. The requirements of China are probably of a similar order. These amounts are more than twenty times the quantities being used today.

Similar improvements can be made in the case of livestock production by the use of better breeds, better feeding practices, disease control, and better management in general. The yield of calories per head of livestock is three times as high in northwestern Europe as in Africa and certain parts of Asia.

Development and Use of Land

Land is the basic resource in food production. In some parts of the world the area of cultivated land is less than one half acre per head of population, and this is decreasing as population rises. At present only about 7 percent of the land surface of the globe is cultivated. Much of the rest is unfit for cultivation by present methods, but there are large areas that could be opened up if capital were available for their development by modern technical methods, including, in many cases, irrigation and drainage. The malaria-carrying mosquito and the tsetse fly now successfully occupy extensive territories that might be devoted to farming if these deadly pests were exterminated by well-organized, large-scale campaigns.

But the opening up of new land is only a part of the problem of increasing land resources. Equally important is the reclamation of land which, once fertile, has been rendered barren by human misuse, and the conservation of land

---

[1]Burns, W. C., *Technological Possibilities of Agricultural Development in India,* Lahore: Government Printing Office, Punjab, 1944.

FIG. 5

INCREASE IN FOOD SUPPLIES REQUIRED BY SOME LESS DEVELOPED COUNTRIES

(IN TERMS OF ORIGINAL CALORIES)

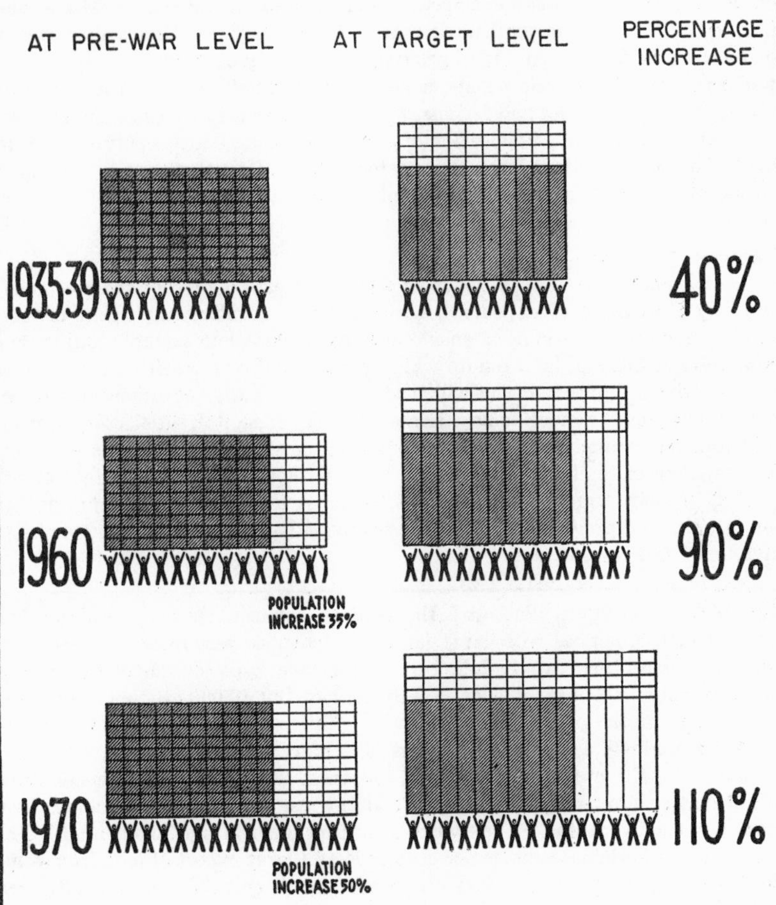

now deteriorating. Every country has its own specific problems in increasing and safeguarding land resources, and they must be attacked in different ways. The main point is that the attack must be scientifically planned and coordinated, and adequately financed by governments.

The Need for Improvement Applies to All Countries

The need for better use of the land applies not only to countries that are at low nutritional levels but to those at medium and high levels as well. In many of the latter a reorientation of production will be required to furnish more of the foods needed for health—livestock products, fruits, vegetables; and this will result in types of farming that favor good soil management, including long-term rotations and increased use of pasture. Indeed, although the discussion in this chapter concerns the less developed countries in particular, much of it applies universally. Studies in the United States of America, for example, show that in many areas fertilizers could be profitably used in greater quantities. Great, well-organized area developments, such as that in the Tennessee Valley, can be useful in many countries and regions. There is no country that has yet achieved all it can achieve by the application of technical advances to agriculture. Moreover, the need, discussed later, for greater industrial development to take more people off the land exists in Europe and parts of the Americas as well as in the Orient.

Economic and Social Changes

Since food production is the most important aspect of the whole economy and way of living of most peoples, a wide range of economic and social changes will be involved in making extensive improvements. For example, unjust and oppressive systems of land tenure which give the cultivator neither opportunity nor incentive to improve his lot will need to be swept away. Since most methods of increasing food production necessitate an outlay of capital, satisfactory systems for supplying credit to farmers are essential; in most countries they do not exist. The capacity of the farmer to develop his land depends to a large extent on the price of primary agricultural products; he must therefore obtain a fair return for the food he produces, and consumers must have the purchasing power to give him a fair return.

Figure 6. This figure illustrates the relation between resources and production. North America, with large land and technical resources—four acres of cultivated land per caput—follows an extensive type of cultivation which produces only 2500 original calories daily per acre but 10,000 original calories per caput daily for the population as a whole. This reflects a high consumption of livestock products and a small surplus for export. South America, with only 1.5 acres of cultivated land per caput, follows a somewhat more intensive type of cultivation producing 4700 original calories per acre; but this includes livestock fed chiefly on the range, which is not included in cultivated land. The high output of original calories per head reflects a large export of food from some South American countries. Western Europe, with only 0.7 acre of cultivated land per caput, undertakes more intensive farming operations and has a higher output per acre than any other region. Even so, the 5250 calories produced per caput is insufficient and has to be supplemented by large imports of food and feedstuffs. In eastern Asia, with small land and technical resources (0.5 acre of cultivated land per caput), cultivation is also fairly intensive, as shown by the figure of 5500 original calories per acre, but the food output furnishes only 2750 original calories per head and there are virtually no imports.

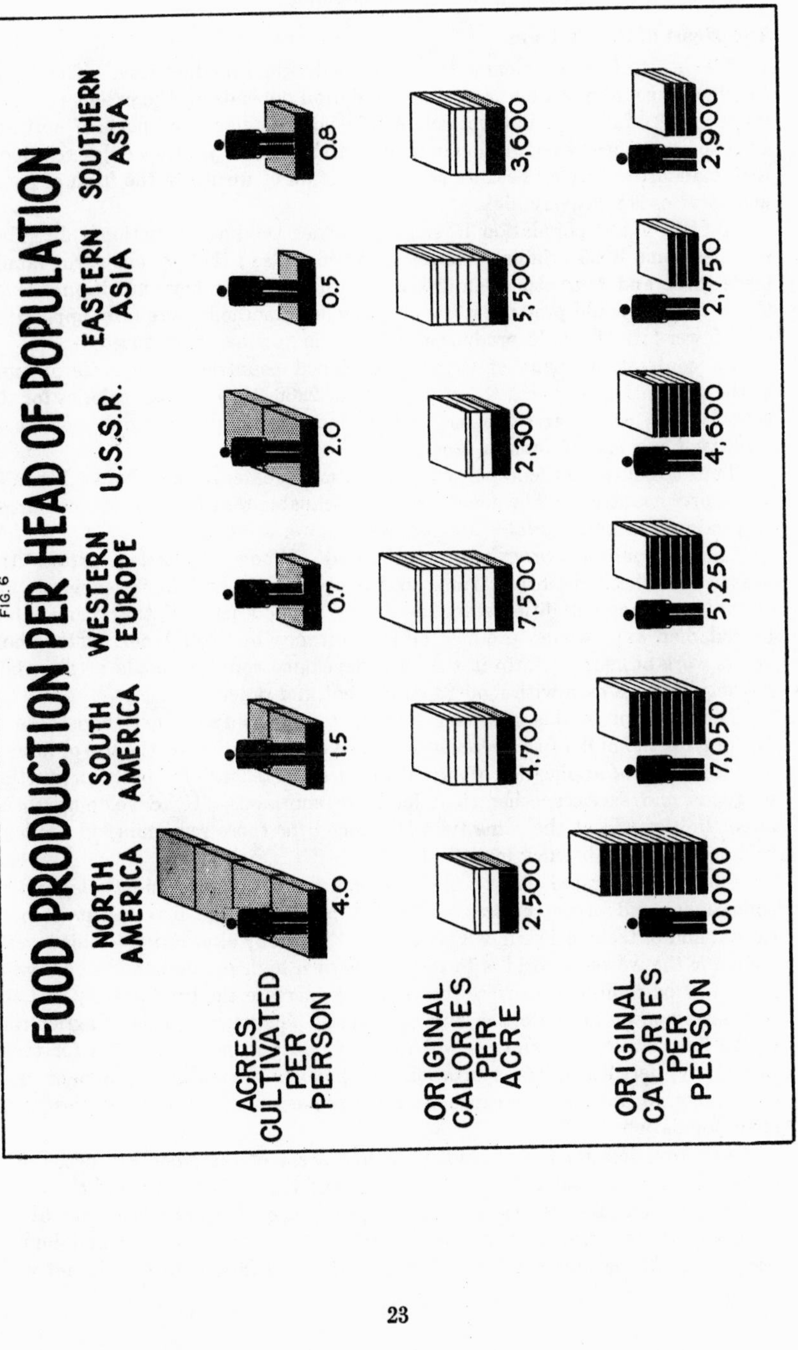
FIG. 6
FOOD PRODUCTION PER HEAD OF POPULATION
NORTH AMERICA
SOUTH AMERICA
WESTERN EUROPE
U.S.S.R.
EASTERN ASIA
SOUTHERN ASIA
ACRES CULTIVATED PER PERSON
4.0
1.5
0.7
2.0
0.5
0.8
ORIGINAL CALORIES PER ACRE
2,500
4,700
7,500
2,300
5,500
3,600
ORIGINAL CALORIES PER PERSON
10,000
7,050
5,250
4,600
2,750
2,900

One of the principal needs in many areas is to wipe out diseases—malaria, for instance—that take a heavy toll of human life, health, and efficiency. Above all, there is the need for education—more and better schools and the development of agricultural extension and advisory services—if farmers are to be able to make full use of modern production methods.

The Heart of the Problem

The heart of the problem is to increase individual productivity. The degree to which a country suffers from overpopulation depends on the extent to which its people are fully and productively employed. England and parts of northern Europe have a heavy population in relation to land area, yet they enjoy relatively high standards of living because their production of wealth in the form of goods and services is relatively high.

A fifth of the population in some countries produces a national diet that supplies some 8000 original calories a person a day; that is, one farm family feeds itself and four other families at a comparatively high nutritional level. These farmers could produce still more if modern methods were fully applied; or even fewer farmers could produce as much as is now being produced.

By contrast, in many of the less developed countries, two-thirds or more of the population produces an inferior diet of 2800–3000 original calories for the country as a whole, and one farm family manages to produce only enough to feed itself and half of another family.

Thus the output of food per man is ten times greater in the advanced than in the poorer countries. The conclusion is inescapable that food for the world can be produced in much greater abundance by fewer hands.

Land resources everywhere are limited. When population presses too heavily on these resources, rural underemployment and inefficiency are inevitable. Human abilities stagnate during a good part of the year. Able-bodied men and women produce only a pittance by their labor. The whole year's work on many a farm in the underdeveloped countries could be done in a few days by one man with modern equipment and practices.

The way out of this situation is to open up resources other than those of farming for the bulk of the population. The opportunities for the use of human skill, through the application of modern science and technology, in the production of goods and services other than food are enormous. By developing them, opportunities will at the same time be opened for those remaining on the land to increase their efficiency manyfold.

This calls for rapid, large-scale development of industry and trade, and of educational and other services. For that purpose, large investment both of capital and of technical skill will be needed. The only alternative to this investment for the western world is to restrict its own high production. The investment will be profitable because it will vastly increase the productivity and the purchasing power of millions of human beings. The improvement of agriculture in the less developed countries will in itself result in large demands for tools, machinery, fertilizers, transportation equipment, processing equipment, and other material, as well as for consumer goods to meet the needs of more prosperous farm populations.

Such advances for great populations and areas of the globe can occur only if the problem is considered a world problem and the challenge a world challenge. The poorer countries cannot master the problem or meet the challenge alone, especially in the present state of development of their material and human resources. All nations will gain by world advances in human health and well-being and in production and trade, and all must participate in bringing them to pass.

Many people who have given serious study to the population problem prophesy doom for much of mankind unless the rate of population growth can be drastically checked. It is worth reiterating that the fundamental solution of the problem lies in increasing the productivity of the individual by putting at his disposal modern scientific knowledge and the tools of modern technology. To the extent that this is done, every individual can become a source of new wealth to his country and to the world. To the extent that it is not done, he is a potential liability, unable to supply his own needs let alone helping to supply those of his fellow human beings.

To put this knowledge and these tools at the disposal of millions of human beings who have never had them requires vision and boldness in the best sense and the highest degree. It must be emphasized that half-way measures will not do. If they are the best that can be devised, the situation will become more and more hopeless, and the prophesies of doom will come to pass. A little amelioration here, a half-hearted attempt at improvement there will serve in the future, as it has in the past, only to increase the numbers of the poverty-stricken and ignorant. The difficulties in the way of formulating and putting into effect vigorous, concerted measures for industrial and agricultural development that will open up new opportunities for these people are very great, but they must be solved or the world faces a future of universally lower living standards or of wars and revolutions that will force the issue.

Making possible sufficiently rapid and extensive progress in the application of science to human needs is thus the biggest and most essential step in solving the population problem. In western countries population growth has invariably slackened, after a preliminary period of increase, as standards of living have risen. There is no reason to believe that the same process will not take place in those countries in which population is now pressing to the limit of subsistence. With economic and social development, history has shown that a change occurs in the whole attitude toward life, hope in the future replacing hopeless acceptance of hunger and poverty as man's natural lot. A social environment is created in which parents consider it of paramount importance that their children should be well educated, vigorous, and healthy, and should have good prospects in life; and emphasis is placed on the development of services and education that foster the ideal of "a healthy mother and a healthy child."

The enormous achievements of the western nations during the war prove that technical means and intelligence and skill equal to the task of bringing about a great economic expansion are available. What is needed now is adequate international action to do the work, and the will to initiate it.

5. MISTAKES AND OPPORTUNITIES

IT IS clear that the world needs greatly expanded food production but that it will not come automatically. The direction of future advances and the need for positive international action have been discussed so far largely from the standpoint of the less developed countries. Action is equally necessary from the standpoint of the more advanced countries, some of which produce surpluses of certain foods over and above their own needs for the international market. Without adequate international action, not only will the world's requirements for food not be met; there is danger of a regression to the trends of the 1930's, when the most technically advanced agriculture in the world had to repudiate its own progress and restrict production to avoid economic disaster. The choice in the highly developed agricultural regions is not one of standing still and holding on to the great gains made during the war, when production was pushed to unprecedented levels. The choice is between going forward and going backward.

It will be worthwhile here to review briefly some of the main trends of the past and the position today.

The Restrictionist 'Thirties

Under the stimulus of the First World War the agriculture of the new world greatly expanded, and in the ten years following the war production in Europe returned to its former level. Together these two sources of supply could have provided sufficient food to feed at least the entire population of the western world at adequate levels of nutrition. Instead the increased supply of food became an embarrassment and the embarrassment a catastrophe. It proved impossible to get the food consumed. Nations had not the purchasing power, the lower income groups within countries had not the purchasing power, and it seemed easier to restrict the supply than to create purchasing power.

During the same period there was a notable increase in the number of countries supplying foodstuffs for the world market, and simultaneously improvements in storage and transportation. All this should have led to greater stability in world market prices. But the reverse happened. Prices fluctuated more violently than ever before. No international machinery existed by which the shocks of economic warfare could be mitigated and conflicting interests reconciled.

There was substantial technological progress in agriculture and a permanent reduction in the real cost of producing many basic foodstuffs. This should have benefited consumers everywhere. Actually it spelled ruin for farmers in many lands, especially for those who had no opportunity to find other occupations when their products became unsalable. It also drove many countries to protect their producers by shutting the door against imported foodstuffs. At the same time technical progress in industry was proceeding at a pace which could have raised the purchasing power of industrial workers in all countries. Instead the world was plunged into economic depression with unemployment so widespread that consumers had to cut down their purchases of food.

Large amounts of savings were accumulated in many of the more highly developed countries and were available for foreign investment. The less developed countries stood in urgent need of capital, but the prevailing atmosphere of depression made governments reluctant to embark on large-scale projects for agricultural and industrial betterment, and surprisingly little development took place where the need for it was greatest. No expert guidance was available internationally to direct the flow of capital, and the money was often devoted to financing doubtfully productive projects in countries in which needs were less acute.

The policies devised to meet the situation in the 1930's took a variety of forms. Some countries which normally imported moderate quantities of food succeeded in reducing their imports until they became virtually self-sufficient. Some continental European countries halved their imports of wheat and reduced those of beef and veal by two-thirds. They kept farm prices high to benefit their own producers, but at the cost of discouraging agricultural progress and worsening the diets of urban workers. The large-scale food importers, in particular the United Kingdom, helped their own farmers by means of tariffs.

quotas, and subsidies, but on the whole did not reduce food imports below pre-war levels.

Countries which were large exporters of agricultural products were faced in the early 1930's with a drastic shrinkage in markets. They were driven to extraordinary expedients in order to save their economies from collapse. The Netherlands, for instance, introduced an elaborate system of controls over the production of all staple foodstuffs and imposed taxes on food consumption in order to give farmers some compensation for the loss of export markets. Similarly, in Australia the home consumer of food was charged high prices to make up in part for the low prices realizable on world markets. In Brazil, during the twelve years from 1929 to 1941, 75 million bags of coffee, one-third of the total output, had to be burned. In these and many other exporting countries the economic crisis brought a heavy decline in national income. Governments tried to give farmers at least as large a share of that income as formerly. But to do this production had to be curtailed and domestic commerce penalized.

In countries such as the United States of America which depended less on the world market, the spread of unemployment caused as great a fall in demand at home as the shrinkage in world trade did abroad. The United States was driven into policies of restriction and large-scale subsidizing of the farming community.

On the international level the record was little better. The League of Nations convened an economic conference in Geneva in 1927 to promote general economic development and freer international trade. The conference was immediately followed by an increase in tariffs in many countries and the imposition of quota restrictions. A World Economic Conference was held in London in 1933. It met at the depths of the depression when the purchasing power of consumers was greatly reduced everywhere, and the only subject on which it could agree was the desirability of further restriction of production. With regard to agricultural commodities, there were protracted negotiations for an international wheat agreement, but these were without success. There were two sugar agreements—the first the so-called Chadbourne Agreement which led to substantial curtailment of production in participating countries, and a second, broader in membership, which achieved the stabilization of a restricted market on the basis of export quotas. There was also a rubber agreement, which brought the world price of rubber to a "remunerative" level by cutting down production and exports.

Altogether, between 1929 and 1939 the world failed to deal with the situation created by the application of science to agriculture and was unable to absorb the increased food supplies thereby made available. This was due partly to disorganization within the sphere of food production itself and partly to instability and fluctuations in the whole economic system. Solutions were sought by each nation acting on its own account and by attempts to deal with commodities separately. Much of this unilateral action had the effect of worsening the general situation.

Beginnings in a New Direction

Though the general tendency in the 'thirties was towards restriction of output, existing policies were being questioned in some quarters and attempts were being made to solve the problem of poverty and underconsumption in the midst of potential plenty.

After the failure of the World Economic Conference in 1933, a new approach was developed in the international sphere. The science of nutrition had advanced far enough to make it possible to define with some accuracy the kinds of diets needed for health, and it had become clear that the greater part of the world's population was getting far less than good nutrition required. In 1935 the Assembly of the League of Nations authorized the League to report on the effect of improved nutrition upon health and the relation of nutrition to agricultural and economic problems. During the next two years the Mixed Committee on Nutrition of the League of Nations reviewed these problems and urged governments to develop food policies which would improve nutrition, especially in the lower income groups, and simultaneously reduce agricultural surpluses. The initiative of the League met with considerable response. A number of nations established national nutrition organizations to advise their governments on policies of nutritional betterment. But the war intervened before much progress could be achieved.

In a number of countries programs were developed which had the effect of bringing surplus food supplies within the reach of those in need, and of raising nutritional levels. These included the provision of nutritious food, free or at low cost, to "vulnerable" groups in the population. School lunch programs and the distribution of milk in schools were organized in some countries. In the United States of America the food stamp plan, followed by the cotton stamp plan, brought surplus supplies within the reach of the unemployed. Numerous other examples could be given.

Some efforts were also made to reduce barriers to international trade. For example, the United States Government, proceeding under the Trade Agreements Act, negotiated agreements with the most important trading countries which resulted in substantial reductions in tariff schedules.

The War and Its Aftermath

The approach and outbreak of war brought about a reversal of the economic policies of the 1930's. Surpluses disappeared almost overnight. All efforts were devoted to expanding the output not only of munitions but also of foodstuffs. Farmers were wooed by favorable prices, guaranteed markets, and a wide range of other inducements and special facilities. They responded vigorously. In the United Kingdom, as already noted, output of food increased by 70 percent in terms of calories, and in the United States by 30 percent as measured in money value at constant prices. Even in Germany and certain other Nazi-controlled countries production remained until near the end of the war almost at the peacetime level, in spite of the great scarcity of manpower, draught power, fertilizers, etc. There was a leap forward in the technical efficiency of farming in a number of countries. Scientific discoveries and methods were applied more rapidly and effectively than ever before.

Under the stress of war remarkable developments took place in food distribution. In some countries marketing arrangements were rationalized, with the result that a number of steps were eliminated. Distribution was facilitated by large-scale use of such methods as the dehydration of milk, eggs, vegetables, and meat. Available food supplies were used to the best advantage for the consumer, partly by rationing programs which ensured that no member of the community should receive more than his share and partly by price policies which held down retail prices so that everyone could afford to purchase the necessary basic foods. In some countries mothers and children were given prior claims on milk supplies. The most familiar example of an effective wartime food policy is that followed in the United Kingdom where, in spite of scantier supplies, the nutritional level of the lower income groups was raised and the health of the nation, as reflected in its vital statistics, improved. The infant mortality rate reached the lowest point yet attained.

Equally important was the international planning of food production and distribution. The Combined Food Board was set up to allocate supplies at the disposal of the United States, Canada, and the United Kingdom. The United Nations Relief and Rehabilitation Administration was created to undertake the task of relieving war-devastated countries and has been concerned with the large-scale and regulated distribution of food supplies in response to urgent needs. Long-term international food problems were thrust into prominence. In 1943 President Roosevelt summoned the United Nations to a Food and Agriculture Conference at Hot Springs, Virginia, leading to the formation of the Food and Agriculture Organization. More recently the United States Government has put forward proposals for an International Trade Organization which will be concerned with the reduction of trade barriers and the stabilization of commodity markets.

These hopeful developments are at present overshadowed by the worldwide food crisis. With the cessation of hostilities the world food situation rapidly deteriorated. Before the end of the war it had been foreseen that the shortages of fats, meat, dairy products, and sugar would remain acute for a considerable time. Towards the end of 1945 it became apparent that supplies of grain would also be seriously inadequate. Among the causes which led to this shortage were the actual devastation of war, which was particularly damaging during the autumn and spring of 1944–45, the serious dislocation of the world's agricultural economy and trade as a result of six years of war, and the dislocation of the world's transport system to serve war purposes. The culminating factor has been a series of droughts during 1945–46 which seriously reduced harvests in various parts of the world.

It is unlikely that the immediate food crisis will be fully resolved until some time in 1947–48. To deal with it, the International Emergency Food Council was set up to replace the Combined Food Board as a result of the Special Meeting on Urgent Food Problems called by FAO in May 1946.

A New Kind of Planning Needed

For the time being, the problem is still one of stretching every resource in the food-exporting countries to prevent famine and alleviate hunger abroad and to meet the demand that, as in the United States of America, results from relatively high purchasing power at home. But a different set of problems looms ahead.

When the present emergency is over, a few major commodities will no longer be in short supply in relation to purchasing power, or effective market demand. Among these are wheat and sugar. There may well be more than sufficient wheat to meet the effective demand after the 1948 crop is harvested. Moreover, there is considerable scope for increasing production in many of the countries in which wheat is grown. Europe's needs will not greatly increase, and the only possible large outlet for increasing exportable supplies appears to be the Far East. It is not easy to substitute wheat for rice, however, in those parts of the East where the latter is the staple food; and in addition, there may be difficulty in making the necessary financial arrangements for large-scale purchases. Sugar should cease to be scarce by 1948 or 1949. The potential world demand is large, and production could be greatly expanded in many tropical areas; but this will happen only if economic conditions are favorable.

It is likely, therefore, that producers of these and other crops will be faced again with a situation in which, if adjustments are left to take care of themselves, the same violent price fluctuations will occur as have characterized the international market in the past.

The situation is likely to be the opposite in the case of fats, meats, and dairy products. The shortage of fats will probably continue for some time, partly because of the reduction in livestock numbers in Europe, partly because some tropical countries that have been important exporters of vegetable fats are finding it necessary to reduce production and export to meet the food requirements of their own people. In comparison with the prewar per caput supply, the shortage of dairy products and meats is likely to continue for several years. It may be seven to ten years before cattle numbers in Europe return to prewar levels, and the scarcity of concentrated feeds may also continue for some time. Meanwhile, if nutrition levels are to be raised, the question of increasing exports of processed milk from the major dairy-producing countries assumes great importance. Part of the need for milk in the less developed countries, for example, might be met by imports of skim milk powder.

This brings up another problem—that of the means for financing international trade. Lack of foreign exchange may make it difficult for many European countries to import the food they need, and there may be a deliberate tendency to forego a better table in order to use the available exchange for capital goods needed for reconstruction. A somewhat similar situation will face the less developed countries. Large amounts of capital will be needed to build up new industries and increase food production, and this may lead to a policy of self-sufficiency and reduction of food imports over a long period so that all resources may be devoted to the task. Yet these countries will in fact need food in large quantities from other parts of the world and would import it if the terms were favorable.

Thus on the one hand there is danger of a return of unmarketable surpluses of certain agricultural commodities, resulting in a disastrous break in prices which would have widespread repercussions throughout the economy and lead again to heavy pressure for restriction of farm output; and on the other, a need to expand production, as indicated by the target figures in this report, in order to raise levels of health and standards of living throughout the world. Unless positive action is taken, the world will move not in the direction of the goals but away from them, and there may again be shortages like those we face in the present emergency, which could have been largely avoided if adequate international machinery had been available.

To avert the danger and supply the world's needs requires a kind of planning and organizing in the field of production, marketing, and finance which neither producers nor nations acting by themselves can carry out. This was recognized in the setting up of the Food and Agriculture Organization. The next step needed in international action is discussed in another report, *Proposals for a World Food Board,* which is the outcome of this world food survey.

APPENDIX I

WORKING GROUPS WHICH ASSISTED IN THE PREPARATION OF THE WORLD FOOD SURVEY

1. NUTRITION TARGETS

Frank G. Boudreau,
Executive Director,
Milbank Memorial Fund,
New York, N. Y., U. S. A.

John M. Cassels,
Office of International Trade Operations,
Department of Commerce,
Washington, U. S. A.

L. A. Maynard,
School of Nutrition,
Cornell University,
Ithaca, N. Y., U. S. A.

Francisco de P. Miranda,
Director, National Institute of Nutrition,
Department of Public Health,
Mexico City, Mexico

Esther Funnell Phipard,
Bureau of Human Nutrition and Home Economics,
Department of Agriculture,
Washington, U. S. A.

Lydia J. Roberts,
Director of Home Economics,
University of Puerto Rico,
Rio Piedras, Puerto Rico

Hazel K. Stiebeling,
Chief, Bureau of Human Nutrition and Home Economics,
Department of Agriculture,
Washington, U. S. A.

2. PREWAR FOOD CONSUMPTION

Joseph A. Becker,
International Commodities Branch,
Office of Foreign Agricultural Relations,
Department of Agriculture,
Washington, U. S. A.

Charles A. Gibbons,
(formerly) Agricultural Economist,
Bureau of Agricultural Economics,
Department of Agriculture,
Washington, U. S. A.

Werner Klatt,
Senior Statistician,
Ministry of Food,
London, England

H. V. Knight,
Senior Statistician,
Ministry of Food,
London, England

Wilfred Malenbaum,
Special Assistant to the Director,
Office of Intelligence, Coordination, and Liaison,
Department of State,
Washington, U. S. A.

Hollis W. Peter,
Head of Food and Agriculture Section,
International, Functional, and Intelligence Division,
Department of State,
Washington, U. S. A.

O. V. Wells,
Chief, Bureau of Agricultural Economics,
Department of Agriculture,
Washington, U. S. A.

Much valuable assistance was also received from other experts attached to the above and other agencies.

APPENDIX II

COMMENTS ON THE STATISTICAL MATERIAL

Calories and protein from estimated prewar food supplies in 70 countries are shown in diagrammatic form in figure 1, opposite page 24. The data on which the chart is based are given in table 2 in Appendix III. Table 1, Appendix III, shows supplies of the various food groups in kilograms a person a year. In most instances the figures represent the average annual supply of the various food groups for the period 1935–38 or 1935–39, but in some instances it was more convenient to take a single year or some other period of years within the decade. The 70 countries include about 90 percent of the world's population; compilation of data for the remaining areas was found to be impossible.

The food groups shown in the charts and tables are as follows:

Cereals

Wheat, rice, barley, maize, millets, oats, and rye. Whole grains and cereal products. Excludes cereal grains used for brewing or distilling.

Roots and tubers

All starchy roots and tubers and their products. Includes white potatoes, sweetpotatoes, yams, taro, and cassava. Starchy fruits (bananas and plantains) are included under this head in countries in which they are a staple article of diet.

Sugar

Refined cane and beet sugar, maple sugar, palm sugar, molasses, panocha, gur, and honey.

Fats

Visible fats and oils, both animal and vegetable. Includes lard and tallow, marine fats and oils, peanut oil, olive oil, coconut oil, margarine, etc. Butter is classed under this head in the case of countries in which it was consumed in substantial amounts and data were available.

Pulses

Mature dry peas, beans, lentils, chick peas, gram, and various other leguminous seeds. Includes also nuts and coconuts consumed as such, and cocoa and chocolate.

Fruits and vegetables

All fruits and vegetables whether fresh, dried, canned, or preserved in other ways, except potatoes and sweetpotatoes. Bananas and plantains are excluded under the circumstances referred to above.

Meat, fish, and eggs

Includes poultry and edible offal. Excludes lard and other animal tissue fats.

Milk

Includes all milk products except butter in the circumstances noted above.

Wine and beer

Other alcoholic drinks have not been included. Their contribution to total calories is unimportant.

In the case of western European and North American countries, the British Dominions and a few other countries, statistical data in more or less suitable shape was obtainable. In preparing the data for countries for which no recent comprehensive surveys of food supplies were available, use was made of relevant information about food production and consumption contained in official and nonofficial reports. This was fairly adequate for cereal supplies, but other foods presented more difficulties. For these foods, arbitrary consumption estimates were sometimes made, based on dietary surveys or other information about food habits in the country concerned. Thus, estimates of supplies of milk and meat had often to be made by inference from livestock figures, themselves of doubtful accuracy. Fish was another dietary item which proved troublesome to estimate. Even in countries in which the fishing industry is well organized, accurate figures of production are difficult to obtain, and when fishing is carried out by more primitive methods, no statistics of any serious value may be available. This is the case in certain tropical and Far Eastern countries, in which fish is nevertheless an important article of diet. Statistics of fruit and vegetable production are defective in almost all countries.

In a number of instances supplies of certain foods were obviously under-reported; for example, reported figures might be quite out of line with those of adjacent countries in which food habits are closely similar, or a large export of some food might be reported with no record of supplies of the same food for internal consumption. Commonsense adjustments were made in such cases.

The figures quoted in the tables are generally "at the retail level" which is equivalent to "food as purchased." To adjust production figures to this level, somewhat arbitrary allowances were made for losses between source and entry into the retail stage. Losses through "spoilage" may be high in underdeveloped countries in which storage conditions are poor, the climate facilitates rapid deterioration, and insect and rodent pests abound. In such countries, on the other hand, the poverty of the people tends to make household waste almost negligible. The reverse is the case in richer and more highly developed countries: spoilage is less and household waste greater. Again, in some countries there are foods which are widely consumed and may be of considerable nutritional importance but are not recorded in food supply statistics; these include such foods as wild fruits and leaves, and game. Reference must also be made to alcoholic beverages, which are included in the food statistics of certain countries and may make a substantial contribution to total calories, while no record of consumption in other countries is available. Another source of error is the inaccuracy of population statistics in some parts of the world.

In calculating the energy value of the food supplies in different countries, a common set of factors for determining the calorie value of foods consumed must be used. Strictly speaking, only a few foods are common to any large number of countries; examples are refined sugar, pure fat, and a few cereal products. Many foodstuffs, even though called by the same names, differ to some extent in nutritive value from country to country, as well as from area to area within any given country, and sometimes even from season to season during the year. The following procedure was adopted in making the statistical survey:

(a) Published figures of the prewar disappearance of food in terms of calories per caput were used in the case of Australia, Canada, New Zealand, the United Kingdom, and the United States of America.

(b) For other countries the calorie conversion factors which seemed most appropriate were used although their validity for the country in question is not established. For Europe, the Middle East, Africa, and Latin America, the factors used were those advocated by the Combined Working Party (1945). For Asia, the publication *Nutrient Values Suggested for Far-Eastern Foods,* issued in August 1945 by the Foreign Economic Administration of the United States of America, was followed.

(c) The calorie figures are also affected by the method of computing the carbohydrate content—and, consequently, the energy value—of different foods. For Australia, New Zealand, and the United Kingdom, the calorie factors are based on "available" carbohydrates, while for Canada and the United States they are based on "total" carbohydrates. According to the report *Food Consumption Levels in the United States, Canada, and the United Kingdom* (1944), prepared for the Combined Food Board, calorie estimates for Canada and the United States would be 100 and 150 calories higher respectively if calculations were based on "total" rather than "available" calories. The calorie conversion factors put forward in *Nutrient Values Suggested for Far-Eastern Foods* for the unmilled or lightly milled cereals give lower calorie values than the customary factors on the ground that these products are not as completely digested as refined cereals. The carbohydrate conversion factors adopted by the Combined Working Party represent a compromise; for many foods they give lower calorie figures than the factors used in Canada and United States, but higher figures than those used in Australia, New Zealand, and United Kingdom.

It must be strongly emphasized that much of the statistical material is provisional and incomplete and that continued revision will be necessary as further inquiries are made and more information becomes available.

APPENDIX III

TABLE 1

PREWAR FOOD SUPPLIES IN 70 COUNTRIES

(kilograms per head per year at the retail level)

Country	Kilograms per year									
	Total kilograms	Cereals	Roots and tubers	Sugar 1/	Fats 2/	Pulses	Fruits and veg. 3/	Meat	Milk 4/	Wine Beer
North America										
United States	718	90	66	48	20	9	203	88	194	-
Canada	637	94	88	47	19	8	98	79	204	-
British Isles										
Eire	783	126	187	38	16	2	90	57	229	38
United Kingdom	608	96	80	50	21	6	113	86	156	-
Scandinavia										
Denmark	806	93	113	55	26	2	131	77	249	60
Norway	672	126	128	36	25	3	55	69	212	18
Sweden	811	97	120	46	18	3	91	69	341	26
Iceland	631	129	62	46	17	2	21	84	270	-
Finland	740	127	180	27	13	3	65	40	280	5
Central & Western Europe										
Switzerland	869	110	90	38	16	2	150	53	320	90
France	846	129	143	22	13	11	137	67	154	170
Germany	723	118	176	24	22	2	86	65	160	70
Netherlands	699	102	130	34	23	10	120	61	203	16
Austria	722	131	85	24	15	2	115	58	240	52
Belgium	757	124	169	26	17	4	70	58	110	179
Czechoslovakia	654	131	160	25	14	4	75	41	150	54
Southern Europe										
Spain	624	148	100	13	15	19	150	44	70	65
Italy	534	170	36	7	10	23	77	34	88	89
Greece	444	146	14	10	17	12	88	32	75	50
Portugal	590	143	70	8	10	10	140	48	70	91
Cyprus	317	164	19	9	8	19	32	25	41	-
Eastern and SE Europe										
Yugoslavia	570	224	40	5	6	6	110	29	120	30
Rumania	605	216	40	5	6	8	130	27	125	48
Bulgaria	524	226	10	4	7	6	100	26	120	25
Hungary	599	187	99	11	4	6	70	36	152	34
Poland	673	150	275	12	8	3	70	31	120	4
USSR	578	198	174	11	5	11	60	28	91	-
Eastern Asia										
Manchuria	328	203	20	3	3	30	53	10	6	-
Japan	391	162	62	15	2	15	93	33	9	-
China	298	166	42	1	5	27	43	14	-	-
Formosa	432	117	158	7	5	14	93	38	½	-
Korea	278	148	24	1	3	20	53	26	3	-
Indian Peninsula										
Ceylon	301	134	13	13	6	63	40	15	17	-
India	296	138	8	15	3	23	37	8	64	-
Southeast Asia Mainland										
Burma	359	164	28	9	7	12	72	49	18	-
Malaya	364	163	25	9	7	19	101	27	13	-
Siam	382	136	46	6	7	17	101	56	13	-
Indochina	363	144	36	6	7	15	101	41	13	-
Major SE Asia Islands										
Java and Madura	369	124	143	4	2	25	59	8	4	-
Philippines	336	118	93	11	5	12	15	46	36	-

Continued

PREWAR FOOD SUPPLIES IN 70 COUNTRIES-Continued

(kilograms per head per year at the retail level)

Country	Kilograms per year									
	Total kilograms	Cereals	Roots and tubers	Sugar 1/	Fats 2/	Pulses	Fruits and veg. 3/	Meat	Milk 4/	Wine Beer
Middle East										
Turkey	506	170	9	5	3	30	114	20	155	-
Palestine	527	177	16	18	12	9	181	25	89	-
Syria & Lebanon	553	145	9	15	4	10	199	11	160	-
Egypt	315	171	3	8	6	24	58	12	33	-
Iran	346	139	-	6	1	10	93	17	80	-
Iraq	334	129	½	9	1	7	96	18	73	-
Trans-Jordan	388	134	-	10	1	9	151	19	64	-
North Africa										
French Morocco	386	170	8	23	6	2	27	39	103	8
Tunisia	381	136	8	17	9	7	3[illegible]	36	110	25
Algeria	342	140	13	15	10	2	40	30	71	21
West Africa										
French West Africa	549	124	240	1	6	17	51	15	95	-
East Africa										
Kenya-Uganda	619	60	464	6	4	20	25	29	11	-
Madagascar	521	134	246	1	3	6	30	20	79	2
South Africa										
Union of South Africa	400	160	22	20	1	9	60	48	80	-
Central America										
Honduras	406	116	36	22	2	10	154	22	43	1
Costa Rica	402	74	6	41	7	15	182	46	31	-
El Salvador	439	101	58	16	6	9	201	35	13	-
Mexico	389	109	22	18	4	12	100	31	86	7
Caribbean										
Cuba	588	92	234	40	11	20	28	47	116	-
Puerto Rico	440	82	137	34	8	25	75	28	48	3
Dominican Republic	602	69	286	15	5	24	138	29	35	1
South America										
Argentina	717	129	54	32	10	9	88	129	200	66
Uruguay	648	101	49	29	5	9	130	132	159	34
Paraguay	718	58	252	12	8	24	116	120	128	-
Brazil	664	93	197	24	6	22	202	60	58	2
Chile	556	115	87	26	6	9	106	45	95	67
Peru	408	112	180	14	6	26	20	20	28	2
Colombia	598	83	174	9	4	5	146	45	123	9
Oceania										
New Zealand	739	94	61	48	17	3	150	128	238	-
Australia	639	94	48	53	17	2	132	134	159	-

1/ Refined sugar equivalent
2/ Pure fat equivalent
3/ Fresh equivalent
4/ Fluid milk equivalent

TABLE 2

CALORIES AND PROTEIN FROM PREWAR FOOD SUPPLIES IN 70 COUNTRIES

(per head per day at the retail level)

Country	Calories per day										Grams protein per day		
	Total calories	Cereals	Roots and tubers	Sugar	Fats	Pulses	Fruits and veg.	Meat	Milk	Wine Beer	Total protein	Animal	Veg.
North America													
United States	3249	887	139	515	502	105	210	524	367	-	88	50	38
Canada	3109	943	177	518	464	81	118	442	366	-	87	47	40
British Isles													
Eire	3184	1205	333	416	393	14	76	329	389	29	92	41	51
United Kingdom	3005	898	125	465	509	67	87	585	269	-	80	43	37
Scandinavia													
Denmark	3249	859	220	603	642	16	81	404	390	34	76	44	32
Norway	3129	1147	242	391	631	27	40	326	313	12	83	41	42
Sweden	3052	911	214	506	437	24	58	327	559	16	88	54	34
Iceland	2980	1184	119	504	434	10	15	261	453	-	101	63	38
Finland	2950	1240	320	300	324	25	44	223	471	3	80	37	43
Central & Western Europe													
Switzerland	3049	1043	162	418	384	17	105	357	472	91	89	48	41
France	3012	1208	247	247	320	100	67	303	222	298	87	38	49
Germany	2967	1117	317	264	528	17	52	399	227	46	77	34	43
Netherlands	2958	967	232	373	573	104	77	340	283	9	78	37	41
Austria	2933	1245	153	264	360	17	76	398	372	48	79	36	43
Belgium	2885	1192	300	285	429	32	44	340	160	103	77	32	45
Czechoslovakia	2761	1242	288	275	336	33	47	277	228	35	72	25	47
Southern Europe													
Spain	2788	1416	180	143	360	158	86	234	101	110	85	20	65
Italy	2627	1606	63	80	256	158	52	148	108	156	81	19	62
Greece	2523	1375	24	113	434	92	129	128	142	86	65	14	51
Portugal	2461	1357	126	88	240	83	87	221	101	158	74	23	51
Cyprus	2304	1570	34	97	200	158	75	100	70	-	85	14	71
Eastern and SE Europe													
Yugoslavia	2866	2065	72	55	144	50	65	191	177	47	87	18	69
Rumania	2865	1995	72	55	144	66	88	179	185	81	87	19	68
Bulgaria	2831	2125	18	44	168	50	56	168	159	43	90	18	72
Hungary	2815	1796	176	117	107	51	43	220	246	59	89	24	65
Poland	2702	1420	495	132	192	25	43	207	186	2	72	19	53
USSR	2827	1837	309	121	131	90	37	153	149	-	88	17	71

Continued

CALORIES AND PROTEIN FROM PREWAR FOOD SUPPLIES IN 70 COUNTRIES--Continued

(per head per day at the retail level)

Country	Calories per day										Grams protein per day		
	Total calories	Cereals	Roots and tubers	Sugar	Fats	Pulses	Fruits and veg.	Meat	Milk	Wine Beer	Total protein	Animal	Veg.
Eastern Asia													
Manchuria	2557	2002	40	32	72	289	44	66	12	-	88	5	83
Japan	2268	1559	157	163	41	147	83	107	11	-	67	12	55
China	2201	1552	112	14	133	260	35	95	-	-	68	5	63
Formosa	2153	1091	434	77	133	102	74	240	2	-	52	14	38
Korea	1904	1419	52	15	72	191	44	106	5	-	70	15	55
Indian Peninsula													
Ceylon	2167	1289	34	140	142	403	51	63	45	-	51	11	40
India	2021	1306	37	163	71	210	41	37	156	-	56	9	47
Southeast Asia Mainland													
Burma	2349	1589	82	103	184	97	62	203	29	-	73	32	41
Malaya	2337	1577	146	100	178	101	66	119	50	-	57	14	43
Siam	2173	1341	200	69	180	114	66	179	24	-	58	21	37
Indochina	2127	1386	143	63	184	96	66	165	24	-	55	17	38
Major SE Asia Islands													
Java and Madura	2040	1154	490	46	61	178	72	35	4	-	43	4	39
Philippines	2021	1115	264	124	130	79	23	218	68	-	55	25	30
Middle East													
Turkey	2619	1605	16	52	84	266	162	103	331	-	101	26	75
Palestine	2570	1610	30	200	297	75	130	94	134	-	74	16	58
Syria & Lebanon	2394	1368	16	164	98	77	242	56	373	-	77	26	51
Egypt	2199	1590	5	88	142	197	43	57	77	-	69	8	61
Iran	1966	1317	-	66	34	88	190	85	186	-	66	17	49
Iraq	1962	1219	1	99	29	58	286	100	170	-	61	16	45
Trans-Jordan	1909	1257	-	110	35	100	174	84	149	-	64	15	49
North Africa													
French Morocco	2431	1595	13	253	153	14	50	121	217	15	75	26	49
Tunisia	2254	1270	15	186	216	63	118	113	225	48	69	25	44
Algeria	2236	1316	23	168	241	18	174	99	156	41	63	19	44
West Africa													
French West Africa	2311	1150	533	13	163	159	35	57	201	-	68	16	52
East Africa													
Kenya-Uganda	2321	586	1231	63	107	167	20	128	19	-	55	12	43
Madagascar	2293	1275	614	11	64	62	25	105	134	3	50	13	37

Continued

CALORIES AND PROTEIN FROM PREWAR FOOD SUPPLIES IN 70 COUNTRIES—Continued

(Per head per day at the retail level)

Country	Calories per day										Grams protein per day		
	Total calories	Cereals	Roots and tubers	Sugar	Fats	Pulses	Fruits and veg.	Meat	Milk	Wine Beer	Total protein	Animal	Veg.
South Africa													
Union of South Africa	2300	1601	46	220	24	63	55	162	129	-	75	23	52
Central America													
Honduras	2079	1072	81	240	56	91	327	163	48	1	49	10	39
Costa Rica	2014	696	12	447	168	142	200	292	57	-	49	21	28
El Salvador	1944	931	128	180	146	84	200	250	25	-	52	14	38
Mexico	1909	1004	42	197	106	108	96	197	155	4	59	20	39
Caribbean													
Cuba	2918	872	587	438	286	217	30	279	209	-	68	29	39
Puerto Rico	2219	793	306	377	199	194	81	189	77	3	55	17	38
Dominican Republic	2130	660	648	168	118	176	103	204	53	-	54	14	40
South America													
Argentina	3275	1240	104	347	241	84	103	755	290	111	111	63	48
Uruguay	2902	971	113	319	132	84	113	830	283	57	102	63	39
Paraguay	2813	545	659	127	187	224	111	719	241	-	99	60	39
Brazil	2552	857	424	263	138	198	141	435	92	4	73	26	47
Chile	2481	1107	178	283	159	84	112	259	170	129	70	24	46
Peru	2090	1033	392	152	142	222	15	94	36	4	58	8	50
Colombia	1934	768	313	98	93	45	100	279	233	5	62	29	33
Oceania													
New Zealand	3281	894	112	507	424	32	119	819	374	-	96	61	35
Australia	3128	872	93	582	420	28	106	745	282	-	90	59	31

FOOD AND AGRICULTURE ORGANIZATION
OF THE UNITED NATIONS

Second
WORLD FOOD SURVEY

ROME - NOVEMBER 1952

This publication was prepared by the staff of FAO. Special mention should be made of the contribution of two staff members who have now left the Organization: Dr. Howard Tolley and Mr. David Lubbock.

Printed in Italy

TABLE OF CONTENTS

INTRODUCTION

One of FAO's first major accomplishments was the World Food Survey, published in 1946. A few months earlier, FAO had been established as the agency through which governments could work together in the task of enabling people of all countries to have enough of the right kinds of food, and to enjoy adequate standards of living. There was a general awareness that a large proportion of the world's population was insufficiently and improperly nourished, but the facts and figures needed to measure the size of the problem had never been systematically assembled. No broad statistical picture or map existed which could serve as a guide in the campaign against hunger and malnutrition which the Member States of FAO had pledged themselves to undertake.

The World Food Survey provided such a picture for the first time. It was concerned with these fundamental questions: What is the food consumption of the different nations? How does it compare with their needs? Where are the most serious shortages? What kinds of food and what quantities of each are needed to achieve improvement in nutrition throughout the world?

In 1946, the war had just ended; the statistical services of many countries were still disrupted; what inadequate figures there were reflected the confusion arising from the temporary food emergency of the war's aftermath. In these circumstances, the prewar period was the latest for which information on a world-wide basis was available. It was, therefore, chosen as the most convenient point of reference in the first World Food Survey. In the absence of reliable national statistics, many of the prewar figures for food supplies and sometimes population were only rough estimates. Nevertheless, the pioneer survey has served a useful purpose. It presented the prewar situation clearly and compactly; it disclosed the main gaps between actual consumption and nutritional requirements; it called attention to the possibilities existing for closing these gaps. It has helped to chart FAO programs and to measure gains or losses in the world-wide struggle against hunger.

Much has happened since 1946 and a new assessment, which will indicate what has happened in the postwar period, is now needed. It is also necessary to gauge the progress which has been made towards the objectives set up in the earlier Survey, and the prospects for the future. Current statistics from many countries are more readily available than they were in 1946, though they still remain inadequate for a complete appraisal of the world's food supply. It is now, however, possible to give some answer to the important questions then raised. Moreover, food requirements themselves can be estimated somewhat satisfactorily by the application of provisional methods for measuring differences in the energy needs of different peoples in different parts of the world; this is the result of the work of international experts brought together by FAO, who tentatively recommended procedures for assessing calorie requirements, taking into account age, sex, body weight and environmental temperature. Other recent developments have also made it easier to gauge the possibilities of progress in food production. More governments are planning and beginning to carry out programs for developing agricultural production and trade, and for improving their food supplies. FAO itself, in the course of its world-wide activities, has gradually been gathering more knowledge of the possibilities for increasing food production in various countries.

The Second World Food Survey is essentially concerned with the same basic questions as the first, viewed in the light of changes that have occurred in the postwar period, and the greater knowledge now available. It may perhaps be claimed that the added material, which takes account of recent trends and prospects, gives greater depth to the new Survey and makes it a better instrument for charting the areas where more food is most urgently needed and for specifying the kind of foods whose production needs to be most rapidly increased.

The new information gives no ground for complacency. The average food supply per person over large areas of the world, five years after war was over, was still lower than before the war. The proportion of the world's population with inadequate food supplies has grown appreciably larger. World food production has indeed expanded since the end of the war, when it fell to a low point, but much of this achievement represents merely a recovery from wartime devastation and dislocation. Clear signs of any far-reaching changes in the entire scale of food production, essential for the improvement of nutrition on a wide scale, are lacking. Annual increases in food production are barely keeping pace with the increasing population. The intensification of health measures in under-developed countries, in particular the use of new methods for controlling mass diseases such as malaria, is likely to lead to a still more rapid growth in numbers. Further, since the Second World War birth rates have been relatively high in most of the well-developed countries, including those which at present produce surplus food. The whole demographical picture, though still imperfectly understood and interpreted, adds a note of urgency to the task of expanding world food production.

All these facts, taken together, scarcely present an encouraging picture of the current situation. It contrasts disturbingly with the view expressed by the Sixth Session of the FAO Conference that " a well-balanced increase of one to two percent per annum in world production of basic food in excess of population growth... is the minimum necessary to achieve some improvement in nutritional standards." The situation which the survey discloses is a challenge to governments and to FAO through which governments are pledged to work together. Some beginnings have been made towards the integrated planning of food production at national, regional, and international levels. It is obvious, however, that the range and intensity of co-ordinated effort must be enormously increased. Granted this, more rapid progress towards the objectives of FAO is possible, whatever the immediate problems and difficulties.

The Second World Food Survey is presented as a report on progress made thus far, and as a guide to future action. It is incomplete and, in many respects, provisional. But if in some degree it assists national governments, regional and international organizations to formulate plans and programs for more intense and comprehensive action in the future, it will have achieved its purpose.

N. E. Dodd

NORRIS E. DODD
Director - General

CHAPTER I

POSTWAR SITUATION

Many of the present significant patterns of food production, trade, and consumption, if they did not altogether originate from the Second World War, have at any rate taken clearer shape as a result. The low level of food production in the under-developed areas of the world, and the wide disparities between food consumption in these areas and in the more advanced countries have long been recognized as outstandingly serious aspects of the world's food and agricultural situation. The effect of the Second World War was to aggravate these problems acutely. Territories were laid waste not only in Europe, but also in the Far East. Destruction of livestock, farm machinery and farm buildings, storage and processing facilities was on an immense scale; soil reserves and sometimes agricultural manpower were seriously reduced. Most of the prolific fishing grounds were closed and the best craft were converted to war purposes. Important sources of supply and markets were cut off from each other. The immense burden of supplying the Allied Powers with food and other requisites for the war effort fell upon the few areas in which supplies were accessible, especially those in which output could be expanded rapidly. In this way, the main features of the postwar dependence of large areas of the world on the surpluses of North America and Oceania emerged. The history of the world food situation during the postwar years is essentially that of an arduous struggle to increase agricultural and other output all over the world, and to restore some balance in the patterns of production and international trade. The struggle is still continuing, its successful issue obstructed by political disturbances, by repeated foreign exchange crises and by recurrent shortages of raw materials and other means of production.

PRODUCTION

Appendix I gives figures for the area, yield, and production of the world's major food crops, region by region, for immediate and more recent postwar periods and a comparison with prewar. In order to eliminate partially fluctuations due to weather conditions, figures for the early postwar period are shown as an average for the years 1946 and 1947; those for the more recent period as an average for the years 1949, 1950, and 1951.

The figures for 1946-1947 amply illustrate the greatly changed situation resulting from the war; the heavy decline in grain, potato and sugar production in Europe, the fall in rice production in the Far East and the remarkable increase in output in North America not only in grain but also in sugar. Changes in the area under crops, though important, contributed only in part to the altered pattern of production. In the world as a whole, the change in the yield per hectare was the major factor. In Europe, the cumulative wartime shortage of fertilizers depleted soil reserves to the point of exhaustion, sharply reducing yields. In Asia the destruction of livestock affected the supply of draft power. On the other hand, on the American continent and to a lesser extent in Oceania, increased mechanization and the rapid improvement in techniques under the stimulus of war were responsible for a rise in yields. The pattern of livestock production emerging from the war was broadly similar to that for crops, with heavy losses in cattle, pigs, and sheep, especially in Europe but also in many parts of Asia (See Appendix II). Pigs, which are usually the first kind of livestock to be sacrificed in times of stringency, suffered the heaviest decline. Both in

North and South America, on the other hand, livestock numbers in all the principal categories increased markedly. Of special importance was the rise in livestock production in South America where, during the war years, forced curtailment of wheat exports, especially from Argentina, resulted in increased internal consumption of grain for animal feed.

The supply of livestock products in food deficit areas was affected even more profoundly than that of vegetable foods. The overall food shortage was so severe in these areas that little grain could be spared for feeding livestock. Imported feed concentrates were scarce and costly. Finally, the major part of the increased feed grain and livestock output in the surplus areas had to be retained to supply an increasing population whose demand for meat and other livestock products was steadily expanding. The early postwar shortage of livestock products, especially meat and eggs, was particularly severe in Europe. This shortage was responsible for large and widening margins between prices paid to farmers for grain, and those prevailing for meat and eggs, on the free and black markets. But even in the Far East, where per caput production before the war was extremely small, the decline in the output of livestock products was proportionately heavier than in other foods.

The year or two immediately following the war marked the low point in the world food position. Supplies were the smallest, distribution the most uneven, hunger and outright famine the most widespread. Since 1946/47 the situation for the world as a whole has improved, but progress towards agricultural recovery has been uneven, fitful and inadequate. Almost everywhere agricultural production has lagged behind industrial production, in many countries accelerating the movement of rural population to the cities. Where, as in part of Asia, pressure of population on the land is severe this movement may in the long run ease the problem of underemployment; but in some other countries, for example Australia and Argentina, continued loss of agricultural manpower is adversely affecting agricultural output.

A concise picture of the scale of postwar recovery and progress is obtained by examining the trend in food production per head in the different regions in the world. A convenient way of doing this is to convert major foodstuffs into a common unit such as wheat equivalent. This is done below for eight major crops, the production per head for each region being expressed as an index of prewar production.

TABLE 1. - *Indices of Total and per Caput Production of Food Crops* [1]

Region	Average 1946 and 1947		Average 1949 to 1951	
	Total	Per Caput	Total	Per Caput
	(.................. Prewar = 100	....	....	..................)
Europe	71	68	96	90
N. & C. America	143	124	150	124
South America	106	87	93	72
Far East	93	85	99	87
Near East	103	91	115	95
Africa	110	96	125	105
Oceania	104	94	116	103
World (excl. USSR)	100	91	111	97

[1] Eight principal crops (wheat, rye, barley, oats, maize, rice, sugar(raw), and potatoes) converted to wheat equivalent (calorie based). Although the above table does not take into consideration crops like millets and sorghum, pulses, sweet potatoes and other starchy roots, of importance in certain areas of the Far East, Africa and Latin America, nevertheless about 80 percent of the world's food supply in terms of calories comes either directly or indirectly through animal products, from the crops included. It, therefore, gives a useful indication of the trends in food production, region by region, in convenient summary form.

Table 1 furnishes striking evidence that recovery in food production from the low levels after the war has barely kept pace with the increase in population in the Far East and Near East. In Argentina food production has sharply fallen from prewar years, but in other parts of South America it has kept abreast of a rapid increase in population. In many parts of the world, per caput food production still remains below prewar levels. One or two countries, like the United Kingdom and Japan, normally dependent to a large extent on food imports, have managed through strenuous efforts to increase per caput domestic food production. But the pattern for most countries within each region is much the same as that for the region as a whole. As might have been expected, recovery was generally greatest in the regions which had suffered the steepest declines during the war. Aided by more abundant fertilizers, increased supplies of farm machinery and other agricultural requisites and by the means to purchase these things through the Marshall Plan, Europe was able to make the most impressive recovery. There were, however, striking differences in the rates of progress in different parts of Europe.

TABLE 2. - *Indices of Gross Agricultural Production in OEEC Countries* [1]

Period	Total Output for Human Consumption	Livestock Products only
Prewar	100	100
1947/48	81	73
1948/49	95	87
1949/50	104	101
1950/51	111	108

[1] Austria, Belgium-Luxembourg, Denmark, France, Western Germany, Greece, Ireland, Italy, Netherlands, Norway, Sweden, Switzerland, Turkey, United Kingdom. Source: *General Statistical Bulletin* published by OEEC, January 1952.

For example, by 1950/51 agricultural production in OEEC countries as a whole was more than 10 percent above prewar and output of livestock products not far behind. Fish production recovered surprisingly fast and by 1950 capacity to produce fish was larger than ever. In Germany and Austria, the division of the countries into different zones of occupation, other territorial changes and many political and economic factors delayed recovery. In the last few years, however, agricultural production in Western Germany has risen with remarkable rapidity but in Austria agricultural output by 1950/51 was still more than 10 percent below and livestock output nearly 20 percent below prewar. Food production in Eastern Europe immediately after the war was much lower than in Western Europe. Recovery in most of the countries in Eastern Europe also lagged behind those in Western Europe. Conditions of chronic food shortage in countries of southern and eastern Europe were of frequent occurrence. For Europe as a whole, the extent of recovery, though large, was little greater than the recovery achieved after the First World War. For example, as shown in Table 3, area, yield, and production of cereals in Continental Europe in the First World War had fallen by much the same order of magnitude as in the Second World War.

TABLE 3. - *Indices of Cereal Production in Europe after the Two World Wars*

Period	Area	Yield	Production
	(.............	1909-13 = 100	.............)
Postwar I [1]			
1920	85.3	84.0	71.8
1925	95.3	103.8	98.8
	(.............	1934-38 = 100	.............)
Postwar II [2]			
1946 and 1947 average	87.7	78.5	68.8
1949 to 1951 average	92.7	101.5	94.6

[1] Post World War I boundaries excluding USSR. Source: League of Nations, *Agricultural Production in Continental Europe during the 1914-18 War and the Reconstruction Period.*
[2] Post World War II boundaries excluding USSR; Production and Yield expressed in wheat equivalents (calorie basis). Source: FAO, *Yearbook of Food and Agricultural Statistics, Production, 1951.*

Yet by 1925, these declines were either nearly or fully made good. Because of some difference in the area covered and for other reasons, comparisons between the two postwar periods are not strictly valid. Nevertheless, the figures even if only roughly comparable are sufficient to remove any impression that European recovery in cereal production after the Second World War has been phenomenal. Nor, if we turn to livestock, is the picture very different. The losses suffered by livestock in the First World War were not quite as heavy as those in the later war; about 10 percent for cattle, 30 percent for pigs, and 15 percent for sheep. Even so they were serious. Yet by 1925 cattle and pig numbers were nearly back to the levels of 1913, sheep numbers were substantially higher. Considerable allowance must, of course, be made for the wider area more deeply involved in the Second World War, the heavier losses in Europe's physical and financial resources, and many other important considerations. The great advances in agricultural techniques achieved over the past thirty years could not, apparently, fully offset these unfavorable factors.

In the under-developed areas of the world, progress since the early postwar years has been much slower. In the Far East, the whole postwar period has been marked by serious political disturbances which have greatly impeded recovery. Average production of rice in the years 1949 and 1950 was still 2.5 percent below prewar while the estimated increase in population was over 10 percent. Since rice normally provides about 70 percent of the food supply of the Far East (in terms of calories) the continuing shortage is a matter of grave concern, especially to the rice deficit countries. Moreover, the increase in production from the low levels of the immediate postwar years has frequently been brought about by expanding the area under rice at the price of a large reduction in the area sown to industrial crops. The situation has been similar for coarse grains like millets and sorghum, and also for pulses. These are food crops of some importance to the region.

Only in the case of starchy roots, like sweet potatoes and yams, has an appreciable increase been recorded. Even this development, however, has more often reflected the difficult food situation rather than general agricultural improvement. Farmers frequently turn to these crops when faced by conditions threatening hunger, since they can be produced quickly in abundance at relatively low cost. The record postwar output of sweet potatoes in Japan is an example of the way in which starchy roots can serve to alleviate food shortage.

The postwar shortage of food, raw materials, and foreign exhange has stimulated programs for the development of African territories. It has also been increasingly realized that the basis for such development must be a widespread and substantial rise in the living standards of the mass of the native populations. Nevertheless, subsistence foods like maize and root crops have languished. Climatic vicissitudes frequently result in crop failures and acute food shortages. Development programs appear to have had more effect on exports like coffee, cotton and especially fats, which were in very short supply in the postwar years. High prices on the world market may have encouraged concentration on these items, sometimes at the expense of subsistence production. To some extent, Africans engaged partly or wholly in producing cash crops have benefited by increased cash incomes, but except where they are protected by price stabilization schemes and other effective marketing arrangements the profits have largely accrued to the merchants. Indeed African farmers, faced with higher prices for the things they need, are sometimes no better off than before.

In Latin America, apart from rice and one or two other crops, there is little sign of a significant upward trend in yields per hectare in postwar years. In the main, except in Argentina where production is greatly below prewar, increased output has come from expanding the area under cultivation. Fish production, although twice as high as in prewar years, is still far from meeting the needs of the region which is importing considerable quantities of fish. The exploitation of the region's potentialities, however, is gradually increasing. If Argentina is excluded, food production appears to be expanding faster than the growth in population. Nevertheless, for the region as a whole the food supply still remains inadequate. High prices for a number of important minerals and other raw materials and specialized export crops like coffee, cocoa, and sugar have sporadically expanded foreign earnings and intensified demand for machinery and other manufactured and consumer goods. In fact, the postwar history of the region has been characterized by waves of prosperity, followed by depletion of exchange reserves and the repeated imposition of trade controls to achieve equilibrium in the balance of payments. Meanwhile, over the years its food export surplus steadily dwindles.

The high levels of food production, developed during and immediately following the war in the main food surplus regions of the world other than Argentina, continued during the postwar years. The persistently high postwar output in North America stands out in striking contrast to the slow and painful progress in most other areas of the world (See Table 4). The region has responded with great resilience to the expanding food needs of its own increasing population. It has also pursued a policy of maintaining food reserves for exports to other parts of the world where under-production has been a feature during the postwar years. The far-sightedness of this policy, in the face of fears of unsaleable surpluses sometimes expressed in various quarters, has been fully justified by events. It is not difficult to imagine how catastrophic the plight of the world might have been in the absence of these food reserves. At the same time, the remarkable succession of abundant harvests in this region, even though partly due to favorable weather, is eloquent testimony of the efficacy of mechanization, soil conservation, research, and the application of scientific practices in raising the entire scale of food production.

TABLE 4. - *Indices of Agricultural Production in the United States of America, Canada and Australia*[1]

Countries	Average of 1946/47 and 1947/48	Average of 1949/50 and 1950/51
	(......... Prewar = 100	...........)
United States of America	128	131
Canada	119	130
Australia	99	112

[1] Source: FAO, *Yearbook of Food and Agricultural Statistics, Production, 1950* and *1951*.

The above summary does not present a particularly cheerful picture of postwar recovery in the state of food and agriculture in the different regions of the world. Nor is the global picture, when analyzed commodity by commodity, more encouraging. This may be seen from Table 5.

TABLE 5. - *World Output of Commodities per Caput (excl. USSR)*

Commodities	Average of 1946 and 1947	Average of 1949 to 1951
	(......... 1934-38 = 100	.........)
Wheat	93.3	99.4
Rice	87.9	89.9
Rye	61.7	83.2
Barley	88.0	99.5
Oats	90.4	96.1
Maize	102.5	106.7
Potatoes	76.7	87.5
Other starchy roots [1]	108.0	110.0
Pulses [2]	88.9	93.9
Sugar	93.3	114.1

[1] Sweet potatoes and yams.

[2] Dry beans, peas, broad beans, chick-peas and lentils.

Despite large postwar harvests in the surplus areas of North America and Oceania, world production per caput has reached prewar levels for only a few major commodities. For reasons already given, output per caput of starchy roots like sweet potatoes, yams and cassava, is higher than of prewar years. The postwar years have also witnessed a phenomenal increase in sugar production. As in the past, many countries continue to aim at self-sufficiency in sugar. Moreover, the need to conserve foreign exchange has stimulated efforts to increase output in Europe and in sterling area territories. Both in Cuba and the Philippines, output has been expanded mainly to meet growing consumption in North America. But these instances are outstanding exceptions. Indeed for some commodities output per caput has been so low as to create acute shortages. This has been specially true of fats. By much effort a considerable expansion in the production of vegetable fats has been achieved, replacing in some measure the heavy decline in the production of animal fats. Nevertheless, even by 1950/51 output per caput of total fats was still below prewar. With the important exception of rice, grain production per caput was gradually restored to levels approaching those of the prewar period. But the devotion of available resources to the production of the staple foods needed by an expanding population left little margin to allow for a major recovery in livestock products. Livestock numbers and output of the world's livestock products per caput of human population, though expanding slowly, remain substantially below prewar.

Competition between the demands for agricultural resources for direct human food and for feeding livestock characterized much of the struggle for recovery in postwar years. This competition was particularly severe in Europe, where programs for expanding the area under coarse grains and fodder crops had to be repeatedly abandoned or postponed to prevent recurrent food shortages from reaching dangerous proportions. Faced with a persistent shortage of feed concentrates, countries were compelled to seek economies by finding alternate and less expensive feeds, by improved silage and by more efficient methods of feeding and of handling grassland. In more recent years, these efforts have had much success, as exemplified by the much higher milk yields per animal than those prevailing before the war and a sharp increase in livestock numbers. Even so, the need to raise the overall production of staple foods above its present inadequate and precarious level severely limits the possibilities of a large expansion in livestock numbers (See Appendix II).

TRADE

The volume of international trade in foodstuffs, though appreciable, has never been large compared with total world food production. From the standpoint of consumption and nutrition, however, such trade can be of considerable importance, since it can spread supplies from surplus food areas to areas where production is below requirements. If the world's growing needs for food and other goods could be met by concentrating production in the areas best suited to produce each commodity at minimum cost, trade in foodstuffs would assume enormous importance. In such ideal circumstances, there could be an increasing interchange of food, raw materials and manufactured products based on economic and efficient exploitation of the world's resources, and on rising standards of living. As it is, neither the volume of trade in food, nor its trend, are necessarily good indicators of the soundness or otherwise of the world food situation. In fact, much of the food trade in the postwar years has merely reflected the distorted patterns of production emerging from the war. Exceptionally heavy exports of food from some areas from time to time have not been signs of general progress. On the contrary, they have usually meant that in other areas of the world there has been a lamentable failure to achieve much-needed increases in production either for domestic requirements or for those of neighboring countries.

Except for bread grain, the volume of international trade in each of the main food groups in 1949-50 was still substantially lower than prewar, especially in rice, coarse grains, meat, and fats (See Table 6). Moreover, apart from bread grains, the volume of exports as a percentage of world production has in each case declined, markedly so in the case of rice.

TABLE 6. - *World Production and Exports of Selected Foods (Excl. USSR)*

Commodity	Prewar			1949-50 (Average)		
	Production	Exports	Percentage	Production	Exports	Percentage
	(...............Million metric tons..................)					
Wheat and rye	149.7	17.3	12	163.2	23.9	15
Coarse grains [1]	196.0	14.0	7	226.2	10.1	4
Rice (paddy)	151.2	14.8	10	152.2	6.5	4
Sugar	26.5	9.8	37	32.9	10.7	33
Fat	21.4	6.1	29	21.5	5.3	25
Vegetable	(11.7)	(4.6)	39	(11.9)	(3.8)	32
Animal and marine	(9.7)	(1.5)	15	(9.6)	(1.5)	16

[1] Coarse grains comprise barley, oats, maize, millets and sorghum.

These declines have taken place not because food deficit countries have increased their per caput food production, though this may be true of one or two countries like the United Kingdom. On the contrary, they reflect principally the inability of deficit countries to buy enough for their food requirements and the failure of some exporting areas to achieve larger surpluses. Especially heavy has been the reduction in trade between countries within the same region. This applies particularly to the trade in rice, the bulk of which is normally conducted within the Far East. The slow rate of postwar recovery in rice production and larger domestic retentions in surplus countries, like Thailand, have sharply reduced the quantities available for export.

TABLE 7. - *Rice Trade in the Far East* [1]

	(1 000 metric tons)
Prewar	13 750
1949 to 1951 (Average)	4 900

[1] Milled rice converted into paddy.

Exports of paddy from Far Eastern countries, for example, are still only 4.9 million tons or nearly two-thirds less than prewar (See Table 7). Moreover, reduced supplies brought a steep rise in the price of rice throughout the postwar years, forcing governments to pay costly subsidies to ensure supplies for needy sections of the population. Political disorders, accompanied frequently by dislocated transportation and inadequate storage facilities, contributed to the difficulties of trade. Territorial changes such as the break-up of the Japanese Empire, in which Korea and Formosa were important sources of rice and Manchuria an important source of soybeans, further affected the pattern of trade. Unable to obtain enough supplies of rice and other foods from the usual source, India, Japan, Malaya, Indonesia, and other countries in the Far East were compelled to import large quantities of bread grains mainly from North America, and even rice from Egypt and Brazil. In this way the region as a whole actually became a net importer of food, whereas before the war it was traditionally an exporter.

The shrinkage of trade within Continental Europe has also been large. Before the war, the grain exports of the Danubian countries and of Poland and Czechoslovakia furnished an appreciable part of the requirements of the rest of Europe. Some of these countries also exported important quantities of livestock products. So far, attempts to restore this trade have proved fruitless. Indeed, with the growing integration of the economy of Eastern Europe with that of the Soviet Union, the prospects of an improvement became more remote year by year. Moreover most of the countries in Eastern Europe are committed to a greater diversification of agriculture, with less reliance on heavy grain output for export, and in order to reduce the undesirably high level of cereal consumption to which they were accustomed before the war. It is, indeed, unlikely that the former role of the Danubian countries as a large grain-exporting area will be restored. Again, the division of Germany into two almost independent zones has had far-reaching effects not only on Germany itself, but also on the whole of Europe. The effects of the separation of the industrialized West from the food-producing East were aggravated by a large influx of population from Eastern Germany and from other former German territories into Western Germany. The enhanced food needs of the West German population could no longer be supplied from the bread basket of Eastern Germany. The deficit could be filled only from other sources, chiefly from North America. Thus, Western Germany imported nearly 6 million metric tons of bread grain and coarse grain in 1949, over 3 million metric tons in 1950, and more than 4.5 million metric tons in 1951. Imports into the whole of Germany before the war were only about 2 million metric tons. Moreover, until the great industrial recovery in Western Germany in 1951, the country lacked the means to buy fruit, vegetables, and other foods which it formerly obtained from countries like the Netherlands, Italy, and Denmark.

Superimposed upon the great decline in intra-regional trade in the Far East and Europe, there was also a steady reduction in the food export surpluses in a number of Latin American countries due chiefly to the rising food needs of the increasing population in these countries and the fall in food production in Argentina. The main burden of meeting the postwar food deficits of the world fell, therefore, on the United States, Canada and Oceania (See Table 8). Food deficit countries have striven to reduce this dependence by developing sources of supply in non-dollar areas, especially in dependent territories in Africa. So far, however, this has not been materially effective.

TABLE 8. - *Exports of Grain: United States of America, Canada and Australia* [1]

Country	Prewar	1949-51 (Average)
	(........ Million metric tons	........)
United States of America	2.4	14.0
Canada	5.2	8.1
Australia	2.9	3.5

[1] Including wheat, rye, barley, oats and maize.

CONSUMPTION AND NUTRITION

How have wartime and postwar developments in food production, utilization, and trade affected the average levels and patterns of food consumption and nutrition in the various parts of the world? The answer to this question is attempted below.

National average food supplies available for human consumption, estimated by the Food Balance Sheet method for the prewar period and for a recent postwar period, are shown in Appendix III. The calorie and protein contents of the same food supplies are shown in Appendix IV.

The prewar estimates (mostly 1934-38 averages) used here are similar to those used in the first World Food Survey, but include modifications needed as a result of some improvement in the accuracy of the statistics. While there are substantial changes in the figures for a few individual countries, the broad picture presented by the earlier statistics remains unchanged. Estimates for the postwar period are available for 52 countries. The exact period chosen for individual countries has usually been the one that best reflected the postwar situation, and afforded a reasonable comparison with prewar. In many countries the year 1949/50 has been taken. For a number of others a recent postwar average of three or four years has been chosen, and for a few the years 1948/49 and 1947/48 have been used in the absence of trustworthy data for later periods.

Appendix V shows that for the world as a whole (excluding USSR) average food supplies, measured in calories after allowing for wastage up to the retail level, were 6 percent lower in recent postwar years than in the prewar period. For a number of reasons this decline was not as large as that in per caput food production in many regions of the world. During most of the postwar years, threatened or actual shortages induced many countries to take exceptional measures to conserve the food supplies. Milling extraction rates were high, the admixture of coarse grains like barley, oats, and maize in bread was appreciably increased, products normally limited to industrial uses (such as many varieties of oilseeds) were frequently used for food. By far the largest economy was achieved by the diversion to human consumption of crops normally reserved for feeding stock. Even so the decline has been marked. It is important to note that for some of the worst-fed areas of the world it has been greater than for the world as a whole. This is illustrated in Appendix IV which shows the distribution of calories and protein for about 80 percent of the world's population and by Table 9 which has been computed from the details in Appendix IV. The remainder of the population for which reliable information is not available, lives almost entirely in regions where food supplies are deficient.

TABLE 9. - *Distribution of Population according to National Average Supplies of Calories and Animal Protein*

Calorie Supplies and Animal Protein	Percent of Total Population [1]	
	Prewar	Recent Postwar
Calorie Levels		
Over 2700	30.6	27.8
2700-2200	30.8	12.7
Under 2200	38.6	59.5
Animal Protein Levels		
Over 30 grams	22.1	17.2
30-15 grams	18.9	24.8
Under 15 grams	59.0	58.0

[1] Comprising approximately 80 percent of the world's population.

The prewar figures shown are in broad agreement with those appearing in the first World Food Survey. The position, unsatisfactory as it was then, has markedly deteriorated, despite some gains since the early postwar years. Over most of the Far East, where nearly one-half of the world's population is concentrated, the declines have generally been of the order of 10 percent. Similar reductions have occurred in parts of North Africa and in a number of Near Eastern countries. In most countries of western and northern Europe average calorie supplies have returned to prewar levels, but in parts of southern, eastern and central Europe, serious declines have not been made good. In contrast, the high calorie levels prevailing in North America and Oceania before the war have been continuously maintained and in most of Latin America, especially in the River Plate countries, there has been a steady and substantial improvement (See Appendices IV and V). Thus, not only has there been an appreciable fall in the average calorie supply for the world as a whole but also the large gaps between the better and worse fed nations have widened.

Since calorie content can be considered as a measure of the quantity of a diet, it would be useful here to consider the adequacy of national average food supplies in relation to estimated physiological requirements. For this purpose, a convenient yardstick in the form of a tentative method of assessing energy requirements is now available (See *Report of the Committee on Calorie Requirements*, FAO, June 1950). This method takes into account environmental temperature, body weights and the distribution by age and sex of a population. It is, therefore, more suited for providing estimates of the average calorie requirements of different population groups than any uniform standard applied to the whole world. For particulars about its application see Appendix III of the above-mentioned report. First, in estimating mean temperatures for the countries, use has been made of publications which give records of environmental temperatures at different seasons of the year, but only for specified cities and other recording stations (See *World Weather Records*, Smithsonian Institution, Washington and *Climate and Man, Yearbook of Agriculture*, U. S. Dept. of Agriculture, Washington, 1941). Considerable uncertainty exists, therefore, as to the appropriateness of the estimated mean temperatures for the whole of each country. Secondly, ranges for body weights were established, representing those within which the weights of a healthy 25 year-old adult " reference " male probably lie for the different regions of the world. The data on which these ranges have been based are scanty and consist, except for a few armed service records, mainly of records of comparatively small numbers of individuals. The top of the range for each region was used in computing the average calorie requirements for individual countries since the weight of a healthy adult male is likely to be above the average of recorded weights of young males in most countries, especially in those countries where food is in short supply. Thirdly, the use of four age groups, 0-14, 15-39, 40-59, and 60 and above, was adopted. Finally an allowance of 15 percent, representing waste up to the retail level, was added to the calculated physiological requirement so that the estimated per caput requirements can be compared with the calorie contents of available food supplies at the retail level. Owing to the tentative nature of the method, the paucity of data and the difficulties involved in estimating body weights and environmental temperatures, the average calorie requirements so computed cannot be more than a rough guide in assessing the adequacy of average consumption levels.

It must be emphasized that the figures for individual countries, estimated by this method, may not correspond closely with those reached by national authorities with fuller experience of the national food situation. Even estimates based on the most complete data available may still, however, be subject to a considerable margin of error. The purpose of including national figures in the table is to illustrate the broad aspects of the world food situation, not to establish definite requirements country by country.

In Table 10, the recent levels of average calorie supplies in some countries for which data are available are compared with the average requirements of their populations. It is immediately evident that the calorie supplies, in general, are short of requirements in all regions, excepting Europe, America, and Oceania. However, wide variations and even some exceptions to the general pattern appear among individual countries in the different regions. Although it is clear that food supplies in many areas are less than estimated requirements, one point should be emphasized. The fact that the average calorie supply in country A is smaller than in country B does not necessarily imply that the former is worse fed, especially when the two countries differ widely as regards environmental conditions and other factors. The physiological requirements of their populations may be different.

The significance of average figures which exceed estimated requirements also calls for special comment. Such excess does not mean that the average person in the countries concerned consumes more food than he or she requires, though there may be certain individuals in the population who are overfed and show a tendency towards overweight or obesity. In well-fed and prosperous countries a considerable " gap " almost always exists between estimated per caput supplies at the retail level and calorie requirements as estimated according to any recognized system, and also between the former and the directly observed consumption of samples of the population. The reasons for this " gap " are not fully understood.

TABLE 10. - *Calorie Supplies Measured Against Requirements*

Region and Country	Recent Level	Estimated Requirements	Difference of Requirements
	(..........Calories	..........)	(..Percent..)
EUROPE			
Belgium-Luxembourg	2770	2620	+ 5.7
Denmark	3160	2750	+ 14.9
France	2770	2550	+ 8.6
Greece	2510	2390	+ 5.0
Italy	2340	2440	— 4.1
Netherlands	2960	2630	+ 12.5
Norway	3140	2850	+ 10.2
Sweden	3120	2840	+ 9.8
Switzerland	3150	2720	+ 15.8
United Kingdom	3100	2650	+ 16.9
USSR	3020	2710	+ 11.4
NORTH AMERICA			
Canada	3060	2710	+ 12.9
United States of America	3130	2640	+ 18.5
LATIN AMERICA			
Argentina	3190	2600	+ 22.7
Brazil	2340	2450	— 4.5
Chile	2360	2640	— 10.6
Colombia	2280	2550	— 10.6
Cuba	2740	2460	+ 11.4
Mexico	2050	2490	— 17.6
Peru	1920	2540	— 24.4
Uruguay	2580	2570	+ 0.4
Venezuela	2160	2440	— 11.5
NEAR EAST			
Cyprus	2470	2510	— 1.6
Egypt	2290	2390	— 4.2
Turkey	2480	2440	+ 1.6
FAR EAST			
Ceylon	1970	2270	— 13.2
India	1700	2250	— 24.4
Japan	2100	2330	— 9.9
Pakistan	2020	2300	— 12.2
Philippines	1960	2230	— 12.1
AFRICA			
French North Africa	1920	2430	— 20.9
Mauritius	2230	2410	— 7.5
Tanganyika	1980	2420	— 18.2
Union of South Africa	2520	2400	+ 5.0
OCEANIA			
Australia	3160	2620	+ 20.6
New Zealand	3250	2670	+ 21.7

Having considered the adequacy of average diets in their quantitative aspects let us turn to the question of their nutritional quality. There is no relatively simple unit like the calorie which can be used for measuring quality, because the latter depends on the presence in satisfactory amounts and proportions of a considerable number of other nutrients, including vitamins and minerals. Protein content is perhaps the best available indicator because most foods rich in protein are also comparatively good sources of many of the other essential nutrients.

Since this is particularly true of foods of animal origin, animal protein level is probably a better indicator than total protein level. It will be seen from both Appendix IV and Table 9 that the quality of food supplies judged by this criterion has been unsatisfactory for the majority of the world's population both before and after the war. Where the food supply is sufficient in calories, it has usually a high protein content, a good proportion of which is derived from animal products. On the other hand, when calorie supplies are inadequate, the total amount of protein in the diet is usually small and supplies of protein from animal products frequently do not reach 10 grams per caput a day. Here too, the position compared with prewar has noticeably worsened, especially in the Far East where consumption of animal protein before the war was already the lowest in the world. Another general indicator of the quality of national food supplies is the proportion of the total calories furnished by cereals, starchy roots and sugar. Where this proportion is unduly high, for example when these foods furnish over two-thirds of the total calorie supply, clear evidence is afforded of nutritional unbalances. In Table 11, relevant data for some countries which are more or less typical of the various regions are shown.

TABLE 11. - *Percentages of Per Caput Calorie Supplies Derived from Cereals, Starchy Roots and Sugar in Recent Postwar Years*

Country and Region	Percent	Country and Region	Percent
EUROPE		NEAR EAST	
France	61	Egypt	79
Greece	66	Turkey	71
Italy	70		
Netherlands	56	FAR EAST	
Norway	53	China	77
United Kingdom	53	India	76
		Indonesia	82
NORTH AMERICA		Japan	82
Canada	47		
United States of America	43	AFRICA	
		French North Africa	77
LATIN AMERICA		Union of South Africa	76
Argentina	57		
Brazil	62	OCEANIA	
Chile	75	Australia	52
Colombia	68	New Zealand	46
Mexico	72		

In almost all parts of the world where the average calorie level has fallen from the prewar average, the high proportion of calories obtained from cereals, starchy roots and sugar has been maintained. In the Far East, where it often exceeds 70 percent, it has tended to increase. The same has occurred in some Near Eastern countries. In the majority of European countries there has been considerable improvement in this respect since the early postwar years, but the proportion is less favorable than before the war. On the other hand, in North America, Latin America and

Oceania, cereals and potatoes now contribute a smaller proportion of the calorie supply than before the war. For sugar alone, however, the proportion has risen. The tendency to consume less cereals and starchy roots and more nutritionally rich protective foods such as meat, milk, eggs, fruits, and vegetables is usually apparent in countries, in which national real income is rising, especially from levels already relatively high; it is significant that North America, Latin America, and Oceania include almost the only countries in the world that emerged from the war with higher real incomes. Elsewhere impoverishment compelled most countries to concentrate on foods of the highest energy content. As available resources have gradually increased, there has been some improvement, but the prewar pattern has by no means been fully restored and per caput consumption of protective foods remains well below prewar levels in many of the under-developed countries (See Appendix III).

Special efforts have been made in many countries to increase the consumption of fluid milk. This has been due partly to a growing realization of the important role of milk in safeguarding the health of nutritionally vulnerable groups, e. g. infants, children, and nursing and expectant mothers. During postwar years, emergency shipments of dried and condensed milk made these products familiar to people not formerly accustomed to milk in this form or, often indeed, in any form. In this period, the volume of international trade in dried and condensed milk has more than doubled; these products have in fact established themselves as important items in the regular international trade in food. The substantial increase in fluid milk consumption in some countries and the large exports of dried and condensed milk has been made possible chiefly by diversion of milk formerly used for making butter, as shown in Table 12.

TABLE 12. - *Milk Consumption and Proportion of Total Milk Used for Butter*

Country	Milk Consumption [1] (Kg. per caput)		Percentage of Total Milk Used for Butter	
	Prewar	Recent	Prewar	Recent
Denmark	195	221	80	70
Norway	207	301	42	30
Switzerland	307	344	20	13
United Kingdom	152	213	15	5
United States of America	249	289	42	28
Australia	165	201	78	66
New Zealand	168	240	67	67

1 Milk in all forms except butter.

In general, the conclusions reached by a study of the national average food supplies are supported by the results of diet surveys which show that the diets consumed by sample groups in most of the heavily populated regions of the world are quantitatively deficient. In plain words, this means that many millions of people do not get enough food to satisfy their hunger. With regard to the nutritional quality of the diets the situation is even more unfavorable because, even in countries in which the calorie levels are adequate, diets often do not contain enough " protective foods." Such unbalanced diets are bound to have serious effects on the health and well-being of the people, a conclusion borne out by surveys of the state of nutrition in various parts of the world which have revealed the prevalence of various nutritional deficiency diseases. Among the most serious deficiency states now prevailing in many areas of the world is a syndrome which displays itself in varying forms and appears to be associated with low protein consumption. It is known to be responsible for high mortality among children from six months to five years of age in Africa, Central America and probably in other regions. Other examples could be quoted to indicate the wide prevalence of malnutrition.

Some comment should be made here on levels and patterns of food consumption of different population groups within individual countries. There is little evidence that in the countries in which average food standards are lower than prewar, the disparities in respect of diet between the well-to-do and poorer sections of the population have grown larger. Governments have become increasingly conscious of their responsibility for safeguarding the health and nutrition of the vulnerable sections of the population. Food rationing and price controls during the war and much of the postwar period, were largely designed to give this protection. Frequently rationed items have included not only bread and cereals, but also protective foods like meat, milk, cheese, and eggs. In addition, many governments assumed a heavy burden of food subsidies to ensure that the supplies of essential foods were within the means of the most needy. As the food situation gradually improved, rationing in most countries was abolished or substantially reduced in scope. Sometimes this was done prematurely and rationing had to be quickly re-imposed when acute shortages threatened to reappear. While rationing and other food controls may have had, in some instances, a deterrent effect on farm production, it is also probable that their abolition has in some countries adversely affected the poorer classes, especially since there has been little pause in the upward trend in food prices. In a number of countries, substantial subsidies on foods are still retained. The provision of supplements and special foods to infants, school children, expectant and nursing mothers and other categories such as heavy manual workers, is also practised on an extensive and growing scale. There is little doubt that in some cases these measures, together with many others, have had important effects. Unfortunately, except in a few countries, little information is available about differences and changes in consumption levels of different groups within the country, which would enable the effect of measures such as those referred to above to be satisfactorily assessed. The only available method of collecting such information is by means of laborious consumption surveys made by careful and statistically valid techniques. The data at present available on differences in food consumption are in general fragmentary and difficult to assemble and interpret.

CHAPTER II

FOOD CONSUMPTION TARGETS FOR 1960

Among the basic ideas which led to the creation of FAO, the concept of food production objectives or targets related to nutritional requirements was of special significance. It received careful thought at the Hot Springs Conference in 1943, and that Conference distinguished two kinds of targets. First, it urged governments to adopt as their ultimate nutritional goals " dietary standards or allowances based upon scientific assessment of the amount and quality of food, in terms of nutrients, which promote health. " Secondly, it drew the attention of governments to the need for more immediate consumption goals " which necessarily must be based upon the practical possibilities of improving the food supplies of their populations. "

THE NATURE OF THE TARGETS

The targets which form the central theme of this survey are of the " more immediate " type. For most countries they do not represent " ultimate nutritional goals, " i. e. food supplies which from the nutritional standpoint are quantitatively and qualitatively adequate for everybody. But these simple statements require considerable expansion in order that the nature of the targets should be clearly understood.

According to its constitution, FAO was established to raise nutritional levels throughout the world. The targets reflect this object, i. e. they represent quantities and patterns of food supplies which, if made available, would improve the nutrition of the people consuming them. It was therefore essential to adopt certain nutritional principles or guiding lines in establishing them; these were approximately similar to those followed in the first World Food Survey. Consideration had, however, also to be given to practicability and feasibility, so that in general the targets represent a compromise between what is nutritionally desirable and hard existing facts and problems. This point will be expanded later.

The nutritional concepts kept in view in setting up the targets may be summarized as follows:

A. *Calories*

(1) For countries in which the calorie value of food supplies at the retail level is below requirements estimated according to provisional methods of the FAO Expert Committee, the target calorie figures in general represent such requirements, plus an allowance of 15 percent to cover losses occurring after retail sale (See previous chapter).

(2) For some countries, the calorie value of existing supplies at the retail level already exceeds estimated requirements. The target calorie figures for these countries were not in any instance set below the existing level. Some adjustments in the upward direction were made to take into account prewar levels and to allow for any unevenness in internal distribution. The procedure adopted was approximately as follows:

(*a*) When the existing level exceeded estimated requirements by less than 10 percent, but was below the prewar level, the target was adjusted approximately to the prewar level.

(*b*) When the excess was from 10 to 15 percent, the target figure was usually set slightly (up to 5 percent) above the existing level, particularly when the prewar level was greater than the latter.

(*c*) When the present level exceeded estimated requirements by more than 15 percent, no change was made even though present levels were lower than prewar levels. In such cases, major consideration was given to improving the nutritional balance of the diet, any proposed increase in certain food groups, such as foods of animal origin, being offset by a reduction in other groups such as cereals and starchy roots.

B. *Protein*

Supplies of total protein (i. e. protein of both vegetable and animal origin) are not likely to be too low when calorie needs are met. No specific minimum figure for total protein was therefore adopted as a guide in establishing the target figures. The protein targets for most countries were, however, adjusted so as to provide for increased supplies of animal protein. This has in general the effect of improving the value of national diets in respect of the quality of the protein they contain and it also improves their nutritive value in general, since most protein-rich foods are good sources of other nutrients. It may be added that the higher the total protein content of food supplies, the less need there is for increasing supplies of animal protein, since a diet already rich in total protein and containing a mixture of different kinds of proteins is likely to fulfill human protein requirements satisfactorily.

With these ideas in mind, the following approach was adopted:

(1) *Animal Protein*

(*a*) When existing supplies of animal protein are less than 25 grams per caput daily, the target figures usually show an increase of 5 grams or more. It may be stressed here that in countries in which supplies of animal protein are very low, there is usually a deficiency of calories also. In such circumstances possibilities for improving the quality of the diet are severely limited.

(*b*) When the present level is between 25 and 40 grams per caput daily, the target figures generally show an increase of between 5 and 10 grams.

(*c*) When the present level is between 40 and 50 grams, no increases were introduced.

(2) *Vegetable Protein*

In considering protein targets, special attention was given to pulses since they have a higher content of protein than other vegetable foods and their proteins supplement those of cereals. Substantial increases in supplies of pulses were included when the level of intake of animal protein is low.

(*a*) When the target for animal protein is less than 15 grams, that for pulses was set between 5 and 10 grams. Even larger amounts were considered appropriate when existing supplies are seriously deficient in calories and animal protein.

(*b*) When the target for animal protein is above 15 grams, it was not always considered essential to set the target for pulse protein above the existing level.

C. *Other Nutrients*

The general pattern of the targets is such as to increase supplies of most important nutrients other than protein, i. e. vitamins and mineral salts which are necessary for health.

D. *Food Groups*

The targets are computed not only in terms of calories, total protein, animal protein, and pulse protein, but also in terms of the major food groups, the figures for the latter reflecting the nutritional concepts referred to in the previous section. The food groups included are those for which estimates of existing supplies are available. The target figures, like the supply estimates, are expressed at the retail level. The following points were among those taken into account in establishing the figures.

(1) *Products of Animal Origin*

(*a*) The targets for the different foods in this group, namely milk, meat, fish, and eggs, depend upon the animal protein target. Potentialities in each country will largely determine which of these foods should make the major contribution towards meeting the animal protein target and the appropriate quantities of each.

(*b*) In designing the target pattern, special attention was given to the need for increasing supplies of milk, because of the special importance of this food in maternal and child nutrition.

(2) *Pulses (including Nuts)*

The targets for pulses are related to the vegetable protein targets. In some national targets in which total calories remained relatively low after the figures for cereals and other food groups were increased to what appeared a reasonable extent, the target for pulses was set at a relatively high figure. Pulses assume special importance when the possibility of raising levels of animal protein intake is not good and present total protein supplies are low. It must, however, be observed that increased consumption of certain pulses, such as soybeans and groundnuts, is practicable only if they are prepared properly and special attention is given to their popularization through educational campaigns.

(3) *Vegetables*

It is safe to assume that in most countries vegetable consumption could with advantage be increased, whatever the present estimated level. Current statistics for this food group are probably the most unsatisfactory of all kinds of food supply statistics and were largely disregarded in setting up the targets. The target figures, expressed per caput per year, range mostly from 60 to 90 kilograms. These high figures are in tune with the potentialities for rapidly increasing production which exist in many countries.

(4) *Fruit*

From the nutritional standpoint, vegetables can replace fruit to a large extent. There is wide variation in present and potential fruit production in different countries and regions. For these reasons, it is not easy to put forward satisfactory target figures. The figures as they stand are in general based on the assumption that the target should at least approach the present or prewar level and that desirable daily per caput consumption is of the order of 150-200 grams.

(5) *Cereals*

In some countries adjustments in supplies of the above food groups along the lines indicated served to bring calorie levels into line with the objectives stated above under A (1). When this was not so and when cereal supplies are not already excessively high, the target figures call for an increase in this food group.

(6) *Starchy Roots*

This food group is mainly of value as a source of energy and is poor in protein and most other nutrients. In general the targets for starchy roots have been set so as not to exceed existing supplies unless there seemed to be no other feasible method of fulfilling calorie needs. Special consideration was, however, given to the advantage of increasing supplies of potatoes and sweet potatoes where potentialities for increased production are large and they are already a familiar and acceptable article of diet. The principle was adopted that supplies of manioc - a root particularly poor in protein - should not be raised, any additional calories needed being obtained from other food groups.

(7) *Fats and Oils*

While no minimum requirement for fat has been established by nutritional science, its value in improving the palatability and acceptability of diets is recognized. Certain animal fats are vehicles of fat-soluble vitamins. Consumption levels in different countries and regions are, however, to a considerable extent a matter of habit.

The method followed in establishing the target figures was to restore the prewar consumption levels when the present level is lower. When both were low, e.g. 20 grams per caput daily or less, the targets usually call for an increase. In some countries in which the consumption of fat is particularly high and total calorie levels are also high, fat targets were set below existing levels if adjustments in the nutritional balance of food supplies seemed desirable without increasing the total calorie yield. These various procedures were related to targets for meat, i.e. additional fat which would become available through an increase in supplies of meat was taken into consideration.

(8) *Sugar*

Targets for sugar are set below present levels when present consumption appeared to be exceptionally liberal and it seemed desirable to improve nutritional balance without raising calorie levels. Otherwise the targets approximate to existing consumption levels.

FEASIBILITY OF THE TARGETS

The targets, as has been said, are a compromise between what may be desirable from the standpoint of nutrition and what may be feasible in practice. They are not ideal nutritional goals for 1960, but rather indicate the general direction along which improvement should move. In establishing them account was taken of the present world food situation and of the potentialities for increasing food production in different countries and regions.

But in interpreting the target figures the great complexity of factors influencing the possibility of achievement must be borne in mind. Some targets demand so large an increase in production that their achievement calls for the most determined efforts. Further, these efforts must cover a wide field, including within their scope such measures as the reform of systems of land tenure, provision for agricultural credit, appropriate adjustments in land taxation, the fostering of co-operatives and the development of extension services. In many instances possibilities of attainment will be influenced by price levels, purchasing power and the readiness of people to change consumption habits. It is clearly impossible to consider each article of food in each individual country, and decide on the chance of its production being increased and its distribution improved in the light of all the relevant circumstances, many of which are unpredictable and imponderable. Some of the measures referred to above may be included in this category. It is therefore inevitable that some of the figures shown in the targets should be somewhat out of line with existing trends and existing realities.

National governments are in a better position to assess the influence of the relevant factors than is the staff of an international organization. The establishment and achievement of satisfactory targets present, in fact, a challenge both to national governments and also to FAO which stands ready to help in formulating plans and to further their implementation through the Expanded Technical Assistance Program. Our immediate purpose here is to establish some goal, or series of goals, in the field of food production and to use these, however open to criticism they may be, as a focal point in our analysis of the existing and future world food situation. If the objectives of FAO are to be achieved, changes and adjustments in national and world food supplies of approximately the kind illustrated by the targets will be needed.

No substantial improvement in the world food situation by the year 1960, in the direction indicated by the targets, is likely to take place unless certain conditions are present. In working out the targets, the following assumptions were made:

(*a*) there will be no major world war or other disasters;

(*b*) average climatic conditions will continue to prevail;

(*c*) the volume of international trade will at least not decrease and such trade will continue to have roughly the same relationship to production as at present;

(*d*) national plans and programs to develop food production will be pushed forward vigorously;

(*e*) technical advances in methods of food production and the application of these will be accompanied by simultaneous advances in other fields, e.g. social, educational, economic and administrative;

(*f*) international assistance to under-developed countries, both technical and financial, will continue to increase.

EFFECT OF THE TARGETS

The lack of precision of the targets in relation to practical conditions has been fully admitted. It will nevertheless be useful, for purposes of illustration, to examine their effect on consumption levels in different countries and regions. In Table 13 the target figures for calories are compared with requirements as estimated by the system previously referred to. While in certain regions,

TABLE 13. - *1960 Targets for Calorie Supplies Measured Against Requirements*

Region and Country	1960 Targets	Estimated Requirements	Difference of Requirements
	(.... Calories per caput per diem ....)		(....Percent....)
FAR EAST			
Ceylon	2200	2270	— 3.1
India	2000	2250	— 11.1
Japan	2210	2330	— 5.2
Pakistan	2230	2300	— 3.0
Philippines	2250	2230	+ 0.9
NEAR EAST			
Cyprus	2480	2510	— 1.2
Egypt	2360	2390	— 1.3
Turkey	2580	2440	+ 5.7
AFRICA			
French North Africa	2290	2430	— 5.8
Mauritius	2240	2410	— 7.1
Tanganyika	2230	2420	— 7.8
Union of South Africa	2510	2400	+ 4.6
LATIN AMERICA			
Argentina	3170	2600	+ 21.9
Brazil	2470	2450	+ 0.8
Chile	2600	2640	— 1.5
Colombia	2590	2550	+ 1.6
Cuba	2820	2460	+ 14.5
Mexico	2420	2490	— 2.8
Peru	2350	2540	— 7.5
Uruguay	2720	2570	+ 5.8
Venezuela	2490	2440	— 2.0
EUROPE			
Belgium-Luxembourg	2880	2620	+ 9.9
Bulgaria	2800	2630	+ 6.5
Czechoslovakia	2810	2640	+ 6.4
Denmark	3120	2750	+ 13.5
Finland	3180	2830	+ 12.4
France	2890	2550	+ 13.3
Greece	2634	2390	+ 10.2
Hungary	2730	2650	+ 3.0
Iceland	3240	2800	+ 15.7
Italy	2680	2440	+ 9.8
Netherlands	3030	2630	+ 15.2
Norway	3190	2850	+ 11.9
Poland	2780	2660	+ 4.5
Portugal	2730	2450	+ 11.4
Romania	2680	2650	+ 1.1
Spain	2700	2460	+ 9.7
Sweden	3120	2840	+ 9.8
Switzerland	3120	2720	+ 14.7
United Kingdom	3120	2650	+ 17.7
Yugoslavia	2440	2630	— 7.2
USSR	3060	2710	+ 12.9
NORTH AMERICA			
Canada	3050	2710	+ 12.5
United States of America	3110	2640	+ 17.9
OCEANIA			
Australia	3150	2620	+ 20.2
New Zealand	3180	2670	+ 19.1

such as the Far East and Africa, the targets do not fully reach requirements, the gap would be considerably narrowed. In India, for example, the deficit would be reduced to one half of its present size, while in the Philippines the considerable existing deficit would disappear. Approximately similar changes would occur in other regions (See also Charts A and B).

CHART A - Calorie Content of National Average Food Supplies (Recent)

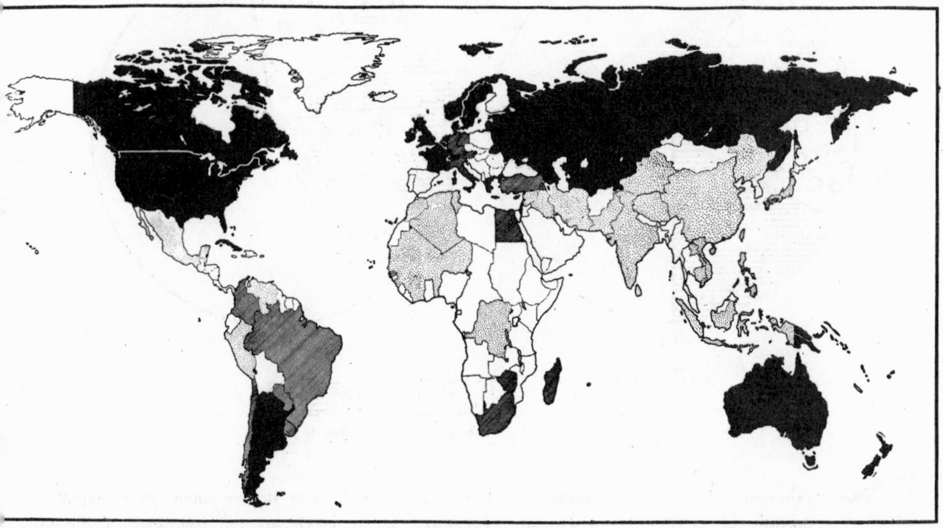

CHART B - Calorie Content of National Average Food Supplies (1960 Target)

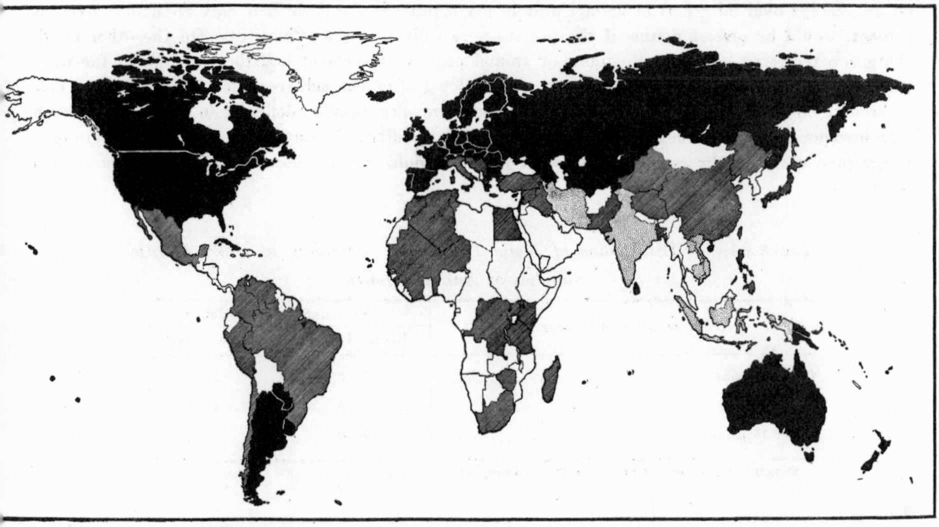

LEGEND

- Over 2,700 calories per caput daily
- Between 2,700 and 2,200 calories per caput daily
- Under 2,200 calories per caput daily
- Data not available

CHART C

Distribution of World's Population According to Daily Average Supplies of Animal Protein
(Comprising approximately 80 percent of the world's population)

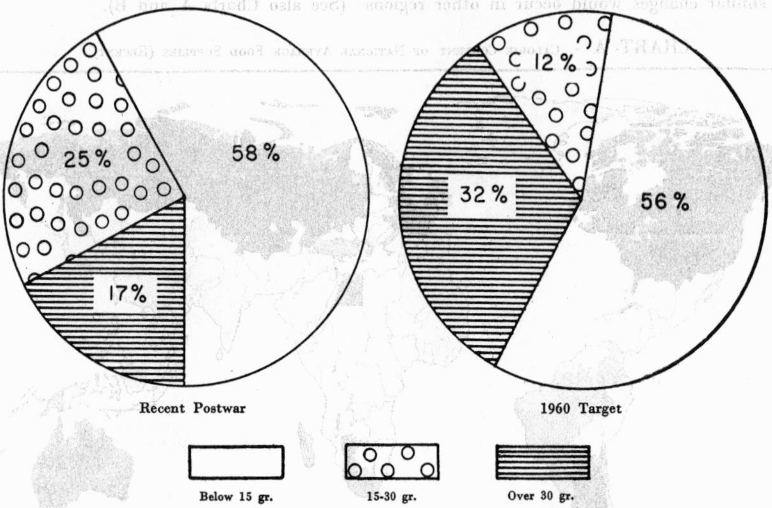

The attainment of the 1960 targets would also lead to definite improvement in the quality of diets, especially in those areas where the prevailing calorie levels are not too low. Table 14, read in conjunction with Table 10, shows that the proportion of the world's population consuming on an average over 30 grams of animal protein per caput per day, which is only about one-sixth at present, would be almost double if the targets were fully met (See Chart C). On the other hand, the group receiving less than 15 grams of animal protein, at present a little over half of the total population, would remain practically at the same level after the achievement of the targets. This is because areas in which animal supplies are low are also usually deficient in calories. In such circumstances, quantity must take precedence over quality. Nevertheless, even in this group a larger precentage of the population than at present would receive more than 10 grams of animal protein.

Table 14. - *Distribution of World Population*[1] *According to Daily Average Supplies of Animal Protein*

Daily Animal Protein Supplies	Percent of Total Population	
	Recent Postwar	1960 Target
Over 30 grams	17.2	32.2
30-15 grams	24.8	12.2
Under 15 grams	50.0	55.6

[1] Comprising approximately 80 percent of the world's population including USSR.

CHAPTER III

ACHIEVING THE TARGETS

THE SIZE OF THE PROBLEM

Appendix VI shows, for each region and subregion and for each major food group, the quantity of gross food supplies required to meet the targets in 1960. The estimates were computed by Food Balance Sheet methods approximately as follows: —

First, the national food supplies needed to provide the estimated 1960 population with average diets set out in Chapter II were calculated. To these estimates were then added allowances for processing and wastage from the stage of production to the retail stage, and for the quantities that would be used for animal feed, for seed, for manufacturing, and other non-food purposes. Many of these allowances are admittedly speculative, since the pattern of crop utilization for feed and non-food purposes in 1960 cannot be accurately predicted. In making them, account was taken both of the present and prewar patterns of utilization, and of the additional supplies needed to meet the targets for livestock products. The estimates of gross supplies do no more than indicate the order of magnitude involved. Theoretically, ignoring changes in stocks, they represent national food production plus imports, less exports. Since for the world as a whole, imports and exports balance, world gross food supplies estimated this way should bear some relation to world food production.

These estimates cover not only the requirements to meet food consumption, set out in Chapter II, but also the additional supplies needed for the increase in population and, in the case of feeding stuffs like cereals and starchy foods, for supporting an increased number of livestock. The figures in Table 15, which summarize the fuller details of Appendix VI, indicate roughly the size of the problem involved if the targets are to be achieved by 1960.

TABLE 15. - ***Percentage Increase Required in Gross Supplies from their Recent Levels to Meet 1960 Targets***

Food Group	Far East	Near East	Africa	Latin America	Europe	N. America & Oceania	USSR	World
Cereals	22	29	33	23	16	16	14	19
Starchy roots	16	66	16	14	21	−13	− 1	12
Pulses	48	61	48	68	56	13	24	47
Sugar	35	16	31	8	6	9	11	15
Fats	...	88	...	...	16	4	32	...
Fruit	...	68	...	...	48	39	39	...
Vegetables	...	14	...	...	39	34	34	...
Meat	47	43	33	28	23	20	20	30
Eggs	24	88	67	55	30	12	12	39
Fish	68	50	75	26	11	18	18	47
Milk [1]	32	31	41	36	16	25	25	33
Estimated percentage population increase by 1960	10	15	15	18	8	14	16	11

[1] Includes all milk products, except butter, as fresh milk equivalent.

For the world as a whole, with the exception of starchy roots and sugar, the estimated increase in gross supplies needed to attain the targets is far in excess of the estimated increase in population. They are particularly heavy for pulses and livestock products. But this by no means indicates the full extent of the problem. To achieve the targets, it is vital that the largest increase in production should occur in the areas where the need is the greatest. Increased production in the surplus regions cannot possibly furnish the expansion required in food supplies for the world as a whole. At best, as was shown in Chapter I, the surplus areas can provide only a small fraction of the needs of the major deficit regions. These needs must be met almost entirely from their own production. The table above shows that great expansion of food supply is required especially in the Far East, Near East, and Africa. In these regions the increase in the supply of cereals must be twice as large as the expected increase in population; for pulses, meat, milk, eggs, and fish, the increase must be proportionately still greater. In Latin America and Europe, except for pulses, the increases called for are smaller, but here too, the estimates for livestock products call for an expansion substantially greater than the estimated population increase. Even in North America and Oceania, livestock production must keep well abreast of the rise in population if the targets are to be achieved. It has been said that adequate data for a reliable estimate of current fruit and vegetable supplies are not available for many countries and regions. Such data as exist, however, indicate the need for a great increase in the supply of these important food groups.

The targets, as pointed out in Chapter II, do not represent the full satisfaction of nutritional requirements. If they did, the increases called for in the supply of many foods, especially livestock products, would be much larger than those shown in Table 15, and beyond doubt, far in excess of what could be achieved by 1960 under the most favorable conditions.

TRENDS IN RELATION TO TARGETS

It is clear from Table 15 that the achievement of the 1960 targets would present a task of formidable proportions. Do trends and prospects suggest that the present efforts to expand production are sufficient to increase the right kinds of food in the right places? Or do targets of this nature imply an effort on a scale far beyond what is at present being put forth or envisaged? The question cannot be answered merely by reference to statistical evidence. First, satisfactory progress is difficult unless political and economic circumstances are favorable and the under-developed countries are provided with adequate financial and technical assistance. There is obviously no assurance that these conditions will be fulfilled during the remaining years of the present decade. Secondly, the trends in food production in recent years largely reflect recovery of ground lost through wartime damage and dislocation. In most areas this kind of recovery can be accomplished more rapidly than continuing advances after the losses of war have been made good. For the most part recovery has now been achieved. Gains in future years cannot be expected to accumulate at the same rate unless special efforts are made. In assessing the prospects for the future from the achievements of the past, the clear distinction between recovery and the solid continuing progress needed to attain satisfactory targets by 1960, must not be overlooked. Viewed in this light the prospects are not at first sight promising. As shown in Chapter I, food consumption for the world as a whole, five years after the war was over, remained appreciably lower than before the war, and substantially poorer both in quantity and quality in the regions in which food supplies were previously most deficient. As far as these regions are concerned, the future endeavor to raise the scale of food production must start at a point lower than that attained in the prewar period.

For the world as a whole, meanwhile, the population is increasing by slightly more than 1 percent per annum. The rate of increase varies considerably among regions. In Latin America, the Near East, and Africa it is about 2 percent; in the Far East it is probably greater than the 1 percent indicated by inadequate statistics; in Europe it is somewhat less than 1 percent. The largest increases in population are still in general taking place in the areas where food supply is most seriously inadequate. It is precisely in these areas that postwar recovery has been least impressive when compared with the increases in food needed to meet the 1960 targets. In the case of cereals, this is illustrated by Table 16.

TABLE 16. - *Annual Rate of Increase in Cereal Supplies*[1]

Region	Postwar Recovery Rate [2]	Approximate Rate of Increase in Gross Supplies Needed for 1960 Targets
	(.................... Percent	....................)
Far East	1	2
Near East......................	2	3
Africa	2	3
Latin America (excl. Argentina) ...	3	3

1 Including wheat, rice, rye, barley, oats, and maize in wheat equivalents. Millets and sorghum are of considerable importance in the Far East and Africa. Full statistics for these crops are not available but partial estimates indicate that their inclusion would not materially affect the above table.

2 Postwar recovery rates were calculated by comparing the average production for the three years 1949, 1950 and 1951 with the average production for the two years 1946 and 1947.

Assuming that cereal supplies from these areas must come almost entirely from domestic production, the figures in the second column also indicate roughly the expansion in cereal output required to achieve the 1960 targets. It will be seen that for none of these regions have the postwar rates of recovery surpassed the annual rates of increase which the attainment of the targets call for. Indeed, for the Far East, Near East, and Africa the rates are only about one-half or two-thirds of those needed. Adequate statistics of pulse production are lacking, but such as there are indicate that the postwar performance in relation to 1960 targets is markedly inferior to that relating to cereals. Among food crops, only in sugar and starchy roots has production in these regions shown signs of outpacing the annual rate of increase called for by the 1960 targets, and these are the food groups which, from a strictly nutritional point of view, should receive least emphasis in programs directed towards a balanced expansion of food production. It can be seen from Appendix I that, except in Latin America, livestock production has failed during the postwar years to keep pace with the increase in population. The extent of this failure is particularly pronounced in the Far East. Yet some of the largest increases needed are in nutritionally important foods like milk, meat and eggs. It is, therefore, evident that even if future gains in these regions continue at the same rate as in postwar years, they would be neither adequate nor oriented in the right direction to meet the targets set out in Chapter II. In most of Europe, chiefly in Western Europe, postwar recovery has greatly outpaced the increase in population, especially as regards sugar, cereals, and potatoes, but also more recently in livestock products. A large part of Europe's food requirements must, however, be met by food imports. Unless such imports are to be heavily expanded, Europe's own food production must continue to climb at a rate far exceeding her growth in population. This applies with special force to livestock products, the consumption of which has still not attained prewar levels. In the circumstances Europe's rapid postwar gains can scarcely be expected to continue automatically.

PROBLEMS AND POSSIBILITIES

With the phase of postwar recovery now over, what factors exist that promise more solid and rapid progress toward our objectives? What are the prospects for a large expansion in agricultural area, especially for food crops? Can fish culture in ponds and rice fields be extended? Do the efforts so far made to increase yields per hectare and per animal by more fertilizers, irrigation, improved methods in animal husbandry and fishing, land reforms and the like portend much higher yields than those of prewar? These are the most important ways by which food production can be substantially increased. What are the prospects that sufficient food exports can be made available to food-deficient countries in a world whose population may have increased by a further 10 percent by 1960? Finally, what prospects exist for bringing about improved nutritional levels through a better use of food supplies? The problems of agriculture, food and nutrition are far too complex to allow detailed answers to be given to such important questions - especially in a survey covering almost the entire world.

Expanding the Crop Area

Appendix I shows that for the world (excluding USSR), the area under major food crops has increased by less than 3 percent since the prewar period. In Europe the area under these crops is lower than prewar, and in Oceania the change is insignificant. The percentage increase has been somewhat larger than the world average in North America, the Far East, Near East, and Africa. Compared with the percentage rise in the world's population, these increases are slight. They scarcely support the possibility that an expansion in crop area can be the major contributor in the attainment of the 1960 targets. There is still, it is true, a large amount of unused but potentially productive land, estimated for the world (excluding USSR) at roughly 370 million hectares, or about one-eighth of the total agricultural area. More of the existing agricultural area, of which only about one-third is arable land, could be put under the plough. Much land at present waste could be claimed for agriculture. (See FAO *Yearbook of Food and Agricultural Statistics, Production, 1950*, pp. 3-7, for further details.) Large-scale programs for reclaiming land which has become almost uninhabitable owing to malaria and other scourges are of some importance in this connection. In the Far East, where pressure of population on land resources is generally most severe, it is estimated that more than 55 million hectares of unused land may eventually be used for agriculture, but these potential resources do not appear to be very great in relation to the existing cultivated area of about 300 million hectares. Moreover, opportunities for individual farmers to bring them into use are apparently extremely limited. The immediate possibilities of achieving the much-needed increase in the scale of production in this region will therefore depend more on raising crop yields than on expanding the area under agriculture. In both the Near East and Latin America, largely unused but potentially productive land could be brought into cultivation given the necessary capital investment. In the Near East alone, these potential resources are estimated at twice the existing arable area. In Africa a similar situation exists, but a vast expansion of knowledge and basic services, especially transport, must precede any rapid increase in food production. In Europe where cultivation is already very intensive, there is little unused land. Scope, however, exists for the ploughing up of grassland, though given the existing situation, it is probable that the area brought into cultivation will be utilized more for high-yielding fodder crops than for crops for direct human consumption.

If a substantial increase in the area under crops is to be achieved in the under-developed and food-deficient areas of the world, an immense amount of capital investment will be required, especially for irrigation and drainage. The last few years have witnessed an impressive number of development programs with this object in view, but although some of these will in due course bear fruit, actual accomplishment has not yet become significant. Much greater effort will be required to promote a more rapid expansion than in the past. However, the financial resources needed for the purpose are a major limiting factor.

Raising Yields

The possibilities for raising crop yields appear to be more favorable. In the more advanced countries, through greater mechanization, more fertilizers and improved techniques, yields per hectare have in many instances strikingly increased over prewar levels. In the United States, where gains have been outstanding, yields per hectare are substantially above prewar levels for nearly all crops. Most remarkable among these gains has been the rise in the yield of maize to nearly double the figure for prewar years, due in part to the widespread and still growing use of hybrid corn and the continued improvement in the quality of the seed. Impressive gains in crop yields over prewar figures have also been registered in Canada and Oceania. In Europe too, even amoung countries where yields were already high because of intensive production, some striking increases have been achieved. Wheat yields, for example, are now 10 percent or more above prewar levels in Belgium, Denmark, Western Germany, the Netherlands and the United Kingdom. No doubt this improvement has been due in part to confining cultivation to more favorable land. In the main, however, it represents the results of using more fertilizers and better techniques.

Contrary, therefore, to widespread beliefs, there is evidently ample scope for increasing yields even in countries where crops are intensively cultivated. If the knowledge and experience in the more advanced countries can be applied elsewhere, with apppropriate modifications, an impressive increase in crop yields could be obtained in regions where the need is the greatest. Especially in the Far East and Africa, but also in the Near East and in some parts of Latin America, yields have stagnated and agricultural productivity remains far below the levels of the advanced countries, frequently as low as one-tenth in terms of output per man-hour. In India, for example, over the past twenty-five years, rice yields per hectare have actually fallen from levels already extremely low. Nor is there any visible evidence of improvement in China during this period. In many countries outside the Far East where rice is produced in appreciable quantities, as for example in Brazil, Egypt, Madagascar, and the United States, yields have risen significantly. The contrasts in yields per hectare in many countries in the Far East themselves provide a further indication of possibilities. Thus, rice yields per hectare in Japan, where the crop is intensively cultivated, are about twice as high as in China and Malaya, and between three or four times as high as in India, Indochina, Burma and the Philippines. The considerations that apply to rice apply equally to other cereals, and to pulses, fruits, and vegetables. An increase of the order of 15 to 20 percent in the yields of these food crops over the next ten years, combined with some increase in the area under cultivation, would bring the 1960 targets for these reasons in sight. It would not call for capital investment on the scale usually needed to bring vast new areas under cultivation. Nor would it imply an extension of mechanization beyond the resources available. Indeed, in most parts of the less developed regions of the world, conditions of cultivation are inappropriate for large-scale mechanization. What is more frequently needed is the provision of simple mechanical tools, the application of more fertilizers, and above all a great reduction in the existing gap between modern knowledge of scientific methods and the primitive agricultural practices which still prevail.

Livestock and Fish

The attainment of a substantial increase in the output of livestock products remains one of the most important longe-term problems, especially in the less developed regions. As pointed out in the first World Food Survey, when crops are fed to animals instead of directly to human beings they lose between 80 and 90 percent of their calorie value before they re-emerge in the form of meat and milk and eggs. This, of course, does not mean that for every calorie produced in the form of livestock products, four or five calories could have been obtained in other forms of human food. Much land used for grazing could not be put to any other effective use; much land at present used for grass or fodder crops could be made to produce food crops, but only of low yield and quality and at very high cost. Rotation of food crops with legume fodder crops is often essential for a sound system of soil conservation. To a very large extent animal and crop production are supplementary rather than competitive, and under the climatic and other conditions prevailing in many parts of the world a system of mixed farming gives best results for both. In some parts of the world animals constitute the main form of draft power in the absence of which crop production would seriously suffer. Nevertheless, there still remains a large field in which livestock can and do consume resources at the expense of human food. This applies especially to the excessive feeding of grain to livestock.

Seen in this perspective the increase in the output of livestock products called for to meet the targets as set out in Table 15 seems formidable. If meat and milk production were doubled or even trebled in most countries of the Far East and Africa and in many countries in the Near East and Latin America, their consumption would still be far short of the present levels in the more advanced countries of the world. However, while carefully avoiding competition by animals for the human food supply, the many existing opportunities for increasing the supply of livestock products should be seized. It is not often realized that some large areas of the world, where consumption of animal protein is very low or negligible, have a far too excessive number of inefficient animals that consume resources and provide little or nothing in return. For example, the Indian subcontinent has the largest cattle population of any area of similar size in the world. Indeed, with only about half of the land area of the United States, it has nearly three times the number of cattle. A similar though less pronounced situation exists in some other countries in the region. In such cases the problem is not one of increasing livestock numbers, but rather of culling herds, improving the efficiency of breeding and feeding and of fighting disease. In many countries the output of livestock products could be at least doubled, without detriment to the production of food crops, by applying better crop rotations, improving grassland unfit for cultivation, using more efficiently crop residues for inedible by-products, and by reducing livestock losses from diseases and parasites. What can be done by improved methods has recently been shown in Western Europe where, with much less reliance on imported feeds than before the war, milk and meat yields per animal have risen steadily by about 2 percent per annum during the past few years, and in a number of countries now appreciably exceed prewar levels.

A more rapid increase in fish production is both possible and desirable, and could do much to compensate for the slowness with which output of livestock products can be expanded. At present only about 10 percent of the world's supply of animal protein comes from fish. The potential resources of the seas and inland waters are, however, considerable. In some countries where programs to increase fish production have been actively pursued, fish consumption has been greatly increased. In Venezuela and Peru, for example, it has more than doubled since before the war. Expanded fish production is an important means of increasing the supply of animal protein at relatively little cost, without encroaching on existing agricultural resources. Although the ocean will be the main source for fish, cultivation in ponds and rice fields offers good prospects, particularly in the Far East.

Reducing Waste

Waste of food between production and consumption still takes place on too large a scale, having regard to the knowledge and means available to prevent it. For example even in the United States grain storage losses due to rats alone are estimated at about 7 million metric tons annually and further losses of between 8 to 16 million metric tons annually are caused by insects. For the world as a whole, the quantity of rice and grain destroyed by rats and insects probably equals the total quantities moving into international trade. Wastage of fish resulting from the present methods of handling and distribution is also substantial. For example at present only about 10 percent of available whale meat is being used for food.

Extension and Demonstration Services

Despite the immensity of the problem, progress is possible if all potentially productive resources — land, farm machinery, fertilizers — can be more fully mobilized. The increasing attention given in the past few years to development programs, research and extension work is an encouraging sign. But the scale of effort remains inadequate. Achievement by governments remains much below that needed if the 1960 targets are to be attained. It has indeed become increasingly clear that development programs, land reforms, the setting up of research stations and training schools, agricultural co-operatives and the like, are only the framework within which expansion in production may be possible. They cannot by themselves assure the accomplishment of the task, unless the individual farmers who are ultimately responsible for food production, are convinced by demonstrations and results of the value of the best techniques appropriate to their circumstances. A tremendous expansion in extension and demonstration work is needed if existing knowledge is to overcome the deep-rooted traditions, prejudices and distrust of farmers whose primitive methods have often remained unchanged for many hundreds of years. For this reason, the Sixth Session of the FAO Conference called on all Member Governments " (*a*) to establish adequate extension and demonstration services which are brought down to the level of the man on the land linked with local administration and education and with the activities of established organizations and institutions and are appropriate to the conditions in their own countries; (*b*) to ensure that the necessary supplies and equipment are available for effective demonstration work; (*c*) to promote where necessary the development of pilot schemes and subsequently demonstration areas in the organizational development of small farmers on a group basis; and (*d*) to provide adequate services to ensure the improvement of home economics in rural areas. "

Greater Incentives

Of equal importance is the provision of proper incentives to stimulate individual farmers and fishermen to increase their output. They obviously will not do this if the increased value of their production is more than offset by high prices for the things they need to raise their output; or if no ready market exists for their surplus production; or if there is no possibility of an improvement in their living standards. In many countries in the less developed regions, the provision of better transport facilities to and from rural districts would enable farmers to sell their products more easily and at better prices. It would also give them greater opportunities to buy more of the goods they need. A growing exchange of goods between farm and city in these regions is indeed essential if a balanced expansion is to be achieved. Among many important incentives to increase production, including subsidies on the use of fertilizers, or for the ploughing up of grassland, assistance in the production and distribution of approved seed, and the like, price incentives are frequently crucial. During the postwar years, wholesale and export prices of foodstuffs and other agricultural products have risen more than those of industrial goods. But in many countries, the full benefit, especially in the case of food crops, has not been passed on to the farmer. On the contrary, the prices he has received have often been fixed at much lower levels. In most countries where this has happened, this policy has been due to the desire to ensure that food prices are low enough to be within the means of the mass of consumers. In a number of countries, food subsidies and similar devices have been employed to protect the consumer and at the same time to give the farmer an adequate price. But many of these devices have proved too costly to be maintained. Frequently, therefore, the farmer has been compelled to accept inadequate prices. Moreover, upward adjustments in the price fixed for his food products have nearly always lagged behind the rise in the prices of industrial goods. In such circumstances, the incentive to increase production is obviously reduced. The situation has been different for industrial crops like cotton, rubber, and wool, whose prices have risen more sharply, due to heavy industrial demand. Because these commodities are valuable earners of foreign exchange, farmers who produce them have as a rule received much more generous treatment. Indeed in some countries a serious diversion of both land and labor from food crops to cotton, rubber and the like appears to be taking place. These trends, if they continue, may further distort the pattern of agricultural production and increase the difficulties of expanding the world's food supply. Flexible price-fixing policies can be of great value in preserving internal stability of price levels and assuring an outlet for production. When food supplies are precarious, however, it is not easy to balance satisfactorily the interests of the producer and the consumer. The encouragement of farm production by incentive prices while at the same time protecting the consumer is a dilemma which so far few countries have been able to solve. Persistently low farm prices, whether fixed in the interest of the consumer or in order to obtain revenue or foreign exchange profits, may easily, as in Argentina, cause a substantial fall in food production, recovery from which may be a long and difficult process.

Expanding Trade

A steady advance towards the higher nutritional levels illustrated by the 1960 targets cannot be achieved if the world is to be faced with recurrent food crises aggravated by shortage of foreign exchange and reduced surpluses in the main food-exporting countries. Whatever the expansion in food production over the coming years, it is probable that a large volume of food imports will continue to be needed by food-deficient countries. Indeed, progressive development of the less advanced areas of the world implies an increasingly effective demand for higher living standards and for food imports into these regions on a scale higher than ever before. Food-deficit countries in the Far East may in future years obtain a larger proportion of their growing food requirements from surplus countries in the same region, but the position of the Far East as a whole as a net importer of food may nevertheless become permanent. Prospects for an expansion in the present low level of trade between Eastern and Western Europe do not appear to be bright. Even if fresh sources of supply are opened up in dependent territories, Europe will still need large food imports from the Western Hemisphere and Oceania. It seems unlikely that the world's dependence on food imports from these two main sources will be substantially reduced in coming years. On the other hand, there is no great assurance that the larger postwar export surpluses will continue. In the United States, steadily increasing domestic demand for livestock products is absorbing more grain for feeding animals. Indeed there is some concern that livestock numbers may have to be reduced unless a further large increase in food grain production can be achieved. During postwar years production in North America has been aided by unusually favorable weather conditions. One or two crop failures might, therefore, cut seriously into the food surplus available for export. In Argentina and Australia, rising domestic consumption has made large inroads into exportable supplies. In both these countries large possibilities of expanded production exist, but increasing industrialization is limiting the degree to which resources are being devoted to agriculture. Unless special efforts are made to maintain the volume of exportable supplies from the traditionally surplus areas, there is a real danger that the requirements of the food-importing countries will not be met.

Nutritional Education

The primary obstacle to the improvement of the diet of the many millions who now suffer from under-and mal-nutrition is unquestionably their low economic status and lack of purchasing power. In a population living at bare subsistence level, the choice of food is severely limited and indeed often non-existent. The foremost need is to satisfy hunger by obtaining enough calories in the form of energy-yielding foods. In such circumstances, real dietary improvement can scarcely take place without economic development. It is only when minimum calorie requirements are satisfied, that serious attention can be given to other aspects of dietary improvement. Where the simple need is for enough food to keep alive, " nutritional balance " is often largely of academic interest.

On the other hand, much can be done to improve the diets of populations living at intermediate economic levels who are not oppressed by extreme poverty. Faulty food habits, based on deep-rooted tradition, prejudice and ignorance are responsible for much malnutrition. For instance, a serious deficiency disease, called beri-beri, is found in Asia and elsewhere among people whose staple food is highly milled white rice, which has been deprived of the essential vitamins because of the mechanical milling of the grain to a high degree. Although the solution of the problem obviously lies in avoiding the use of such rice, there are several obstacles, mainly of a social and psychological character. People accustomed to highly milled white rice do not take kindly to other kinds of rice, such as under-milled and parboiled rices, which are more nutritious but less attractive to the eye and the palate. Many other examples of social or cultural obstacles could be quoted. These may be religious taboos, such as those which prohibit eating meat from the cow or the pig, or they may be individual and collective prejudices, e.g. against the consumption of milk, fish, eggs and so on. Sometimes the taboo or prejudice is not related to the consumption of particular foods but to their production or distribution. For example, raising vegetables is considered as an inferior occupation in some areas and is therefore undertaken only by immigrants.

Another important obstacle to improving diets is the difficulty of popularizing new and unfamiliar foods. General experience indicates that such popularization, while by no means impossible, inevitably takes time. The speed and nature of proposed dietary changes, however desirable these changes might be from the nutritional standpoint, must necessarily be adjusted to prevailing food habits.

THE NEED FOR AN INTEGRATED APPROACH

It is evident that the larger food supplies needed to achieve "intermediate" targets by 1960 and the further increases required to bring average diets closer to full requirements of all countries will not be forthcoming without far greater effort than has so far been put forth. Beginnings have been made, but only the fringe of the problem has been penetrated. In the less developed areas, the vicious circle of poverty, hunger, malnutrition, ill health and physical inefficiency remains unbroken. In a sense, the need for action is more urgent than it has ever been in the past. Unsatisfied demand for food and better living conditions has become increasingly vocal. People are less ready to tolerate hunger and famine. Governments must strive to provide enough food and better standards for their people or face the danger of social upheaval. Effective and sustained action on a broad front and integrated planning at all levels - national, regional and international - is required.

At the national level, measures to raise the output of food by development programs to improve methods of production, etc., must be dovetailed into the national plan for economic betterment. Technical assistance must be directed towards projects likely to yield maximum benefits to the country as a whole. Land reforms must be planned not only to remove injustices that have made progress impossible, but also to ensure that farming efficiency is preserved and increased, and that the transition to new systems of land tenure is made without a disruption of the country's economy. Agricultural programs must be closely linked with nutritional policy to ensure that the right kinds of food as well as enough food are produced. They must be closely co-ordinated with plans for industrial development to attain a balanced expansion in the country's economy, including a growing interchange of products between farm and city. Finally, steady advance in these fields cannot be assured without a parallel advance in many others, including education, health and hygiene, housing, transport and the like.

Integrated planning and action at the regional level is equally important. The spectacular rise in food production in North America over the past few decades was not the result of good luck. On the contrary, in a world with a growing population and a growing multiplicity and complexity of wants, the advantage lies with nations possessing a large territory under unified economic control, where land and other resources are ample in relation to population and where modern techniques can be developed and applied without hindrance. In many of the less developed regions, land and other potential resources are available in abundance, but unified control has been largely absent and modern techniques have not penetrated to the masses. In other parts of the world, as in Europe, some national units have become too small for full advantage to be taken of modern forms of industrial organization and technique. Persisting trade and other barriers have resulted in diminishing returns to a point at which economic standards can scarcely be maintained. The lesson is clear. In such circumstances there must be a greater pooling of resources, technical knowledge and experience within each region. National plans and programs must be co-ordinated within a regional framework to ensure that output is expanded in areas best suited for the purpose. Trade and other restrictions within the region must be reduced to a minimum so that a ready market can be found for increased production. Only in this way can the basis be laid for the more unified economic control so vital for rapid and orderly regional development. Beginnings in this direction have been made but the scope of co-ordinated regional planning and action must be greatly enlarged if really significant results are to be obtained.

On the international level, an integrated approach is no less vital. Expert knowledge available all over the world cannot be of full benefit to the less advanced countries unless the resources of these countries are more efficiently mobilized. Again, development programs frequently require capital investment on a scale far beyond the resources of individual countries. Private organizations and institutions cannot be readily persuaded to lend funds for this purpose, unless part of the risks involved can be transferred to shoulders better able to bear it. International collaboration is needed to ensure the orderly marketing of foodstuffs at reasonably stable prices, fair alike both to the producer and consumer. Steps must be taken to avoid the dangers of recurrent world food shortages. In particular, special efforts may be called for to make certain that sufficient exports will be available from the great food surplus regions of the world to meet the urgent needs of the deficit areas, while these in turn are developing their own resources. International machinery is also necessary to prevent serious local food shortages from developing into famine conditions and for the application of immediate relief when such famines arise.

It was pointed out earlier that the attainment of satisfactory targets by 1960 cannot be envisaged unless political conditions over the next ten years are favorable for orderly progress. Since the outbreak of the Korean war, many countries have embarked on heavy defense programs entailing a large diversion of resources to armaments. This has already resulted in shortages and high prices for agricultural machinery, fertilizers and many other agricultural requisites. It cannot be over-emphasized that freedom from hunger in itself helps towards peace and security. Measures to obtain this freedom are therefore at least as important as defense measures. Any considerable diversion to other uses of the existing precarious resources available to agriculture can easily jeopardize the world's food situation and thus increase the very dangers which have led to the present large-scale program for defense.

Unless the present trends in world population undergo some unforeseen and appreciable change in the immediate future, it is unlikely that current estimates of population for the decade ahead will differ by more than a small percentage from the number of people actually then in existence. If, therefore, a moderate and world-wide improvement in food and nutritional standards is to be achieved within the next ten years or so, food supplies must be increased by an amount of the order of magnitude indicated in this Survey. This does not imply that the effort needed to achieve better food standards may be relaxed at any time in the forseeable future. On the contrary. The 1960 targets by no means represent fully satisfactory nutritional standards. But if they can be achieved, even only in part, within a reasonable time, their maintenance will probably require continued efforts on an increasing scale.

In the industrially developed countries the historic trend of declining birth rates not only came to a halt in the early thirties, but was sharply reversed especially during the years following the world economic depression, and in the early postwar years. This change, which may prove to be temporary, has already been responsible for an unusual upsurge in the population of these countries which is likely to persist at least for some years. In the densely populated, less developed areas, the potential effect on population of the application of the modern techniques of medical science is tremendous. The present absolute annual increase in population in these regions is great, although because death rates as well as birth rates are extremely high, the actual rate of increase in population is moderate, ranging mostly from 1 to 1.5 percent per annum. If the economic pattern in these countries were to follow closely that of the industrially developed countries, improvement in economic standards gradually accompanying industrialization would bring about a slow and steady fall in birth and death rates. The accumulated knowledge available to medical science and the technical facilities at its disposal has, however, made it possible to reduce mortality rates far more rapidly than was possible in the past. Because " mass diseases " like malaria, tuberculosis, etc., affect such a high proportion of these populations, the control of these diseases alone, now possible at relatively little cost, can bring about a striking reduction in mortality. For example, in Ceylon the death rate, recently reduced mainly by successful measures against malaria to nearly one-half of its previous level, now approaches the death rate prevailing in industrialized countries. Similar results are being achieved elsewhere. Since, however, birth rates are still in the main determined by longer term factors relating mostly to economic and social attitudes, they are not susceptible to equally rapid change.

Clearly the planning of food supplies for the future must therefore take into account the possibility of a much more rapid increase in world population. The problem may be eased by the improvement in efficiency that can result from successful health measures. A number of examples have been reported of increased food production achieved entirely through improved efficiency following successful attacks upon disease. The size of the contribution which better health can make towards alleviating the problem in the long run is, however, unpredictable.

In the last few years, consciousness that the world is one has grown apace. There is increasing realization that the better-off countries must assist the comparatively poorer nations, not merely from humanitarian motives, but also to safeguard their own living standards. There is also a greater understanding, which in some degree is already being translated into action, of the need for integrated planning at all levels to achieve higher living standards for people all over the world. But, as this report has emphasized throughout, actual achievement is still all too meagre.

APPENDICES

The tabulation of the data contained in the Appendix tables was completed in January 1952 and many of the figures in the tables, especially in Appendices III, IV, V and VI, were based on estimates obtained in earlier periods. More up-to-date estimates would involve changes in many of the individual figures. However, it is not believed that such changes would significantly affect the order of magnitude of the figures shown. In these Appendices, as in the main tables in the text, the major regions have been arranged in the order of their per caput food supplies, starting with the region in which per caput supplies are the smallest.

APPENDIX I - AREA, YIELD, AND PRODUCTION OF MAJOR CROPS

Commodity	Europe			North & Central America			South America			Far East		
	1934-1938	Average 1946 and 1947	Average 1949-51	1934-1938	Average 1946 and 1947	Average 1949-51	1934-1938	Average 1946 and 1947	Average 1949-51	1934-1938	Average 1946 and 1947	Average 1949-51
	(....AREA - 1,000,000 hectares....)											
Wheat	29.8	25.7	28.3	33.1	39.0	38.3	8.6	7.0	6.7	36.9	40.2	36.6
Rye	13.5	11.1	12.3	1.6	1.1	1.1	0.5	0.8	0.7	–	–	0.1
Barley	9.4	8.5	8.9	5.7	7.3	7.1	0.9	1.3	0.9	11.6	11.2	11.6
Oats	14.6	12.9	12.8	19.6	21.2	20.7	1.0	0.9	0.8	1.3	1.2	1.2
Maize	11.7	11.1	10.8	41.9	39.5	39.5	10.0	8.6	8.1	13.1	13.5	14.7
Rice (paddy)	0.2	0.2	0.3	0.5	1.0	1.1	1.2	2.1	2.4	81.4	80.0	86.6
Potatoes	10.0	9.0	9.3	1.5	1.2	1.0	0.6	0.7	0.8	1.1	1.3	1.3
	(....YIELD - 100 kg. per hectare....)											
Wheat	14.2	11.0	14.7	8.2	11.5	10.9	9,6	11.4	9.8	9.4	8.3	9.8
Rye	14.2	11.0	14.6	7.4	7.6	7.6	6.0	6.6	5.9	6.9	6.4	7.8
Barley	15.3	13.1	16.6	11.1	12.6	13.9	9.4	11.3	11.0	11.0	10.2	10.5
Oats	15.7	13.2	15.6	9.7	11.6	12.6	11.0	9.7	9.6	8.6	7.2	8.3
Maize	14.8	10.8	14.1	13.3	19.1	21.3	15.3	15.9	12.9	11.1	10.5	10.4
Rice (paddy)	51.8	40.9	43.8	21.0	20.6	21.3	15.3	17.1	17.1	17.7	16.8	16.2
Potatoes	134.6	115.2	137.0	78.5	118.3	137.9	43.6	57.9	59.0	73.0	67.9	66.4
	(....PRODUCTION - 1,000,000 metric tons....)											
Wheat	42.3	28.3	41.6	27.0	45.0	41.7	8.2	8.0	6.6	34.8	33.5	35.8
Rye	19.1	12.2	17.9	1.2	0.8	0.9	0.3	0.6	0.4	–	–	–
Barley	14.4	11.2	14.8	6.3	9.2	9.9	0.8	1.4	1.0	12.8	11.5	12.2
Oats	23.0	17.0	20.0	19.0	24.6	26.0	0.9	0.8	0.7	1.2	0.8	1.0
Maize	17.4	12.0	15.2	55.9	75.3	84.0	15.3	13.7	10.5	14.6	14.2	15.2
Rice (paddy)	1.1	0.9	1.3	1.2	2.0	2.4	1.8	3.5	4.1	143.8	134.6	140.2
Potatoes	135.0	104.2	127.9	12.1	14.3	13.1	2.8	4.1	4.8	7.9	8.6	8.9
Sugar beet / Sugar cane	6.5	4.8	8.2	7.0	10.6	11.5	2.5	3.7	4.2	7.3	5.0	6.4

APPENDIX I - AREA, YIELD, AND PRODUCTION OF MAJOR CROPS *(concluded)*

Commodity	Near East			Africa			Oceania			World (Excl. USSR)		
	1934-1938	Average 1946 and 1947	Average 1949-51	1934-1938	Average 1946 and 1947	Average 1949-51	1934-1938	Average 1946 and 1947	Average 1949-51	1934-1938	Average 1946 and 1947	Average 1949-51
	(.......AREA - 1,000,000 hectares.......)											
Wheat	9.5	9.1	11.2	4.8	4.4	5.1	5.3	5.6	4.7	128.0	131.0	130.9
Rye	0.4	0.4	0.5	–	0.1	0.1	–	–	–	16.0	13.5	14.8
Barley	4.3	5.2	5.3	3.6	2.8	3.9	0.2	0.3	0.5	35.7	36.6	38.2
Oats	0.3	0.3	0.3	0.4	0.4	0.4	0.7	0.8	0.8	37.9	37.7	37.0
Maize	1.3	1.6	1.6	6.6	9.1	7.7	0.1	0.1	0.1	84.7	83.5	82.5
Rice (paddy)	0.8	1.0	1.0	1.7	2.3	2.6	–	–	–	85.8	86.6	94.0
Potatoes	0.1	0.1	0.1	0.1	0.1	0.1	0.1	0.1	0.1	13.5	12.5	12.7
	(.......YIELD - 100 kg. per hectare.......)											
Wheat	10.4	10.4	9.3	5.3	5.2	5.9	8.2	8.4	11.1	10.1	10.0	11.0
Rye	9.6	9.5	9.2	4.5	4.7	–	6.7	4.6	–	13.1	10.4	13.2
Barley	9.7	7.6	9.2	5.8	5.9	6.5	9.9	12.1	11.3	11.4	10.8	12.0
Oats	9.6	7.8	9.6	7.4	6.2	7.5	5.5	7.0	7.1	11.9	11.8	13.2
Maize	18.6	14.7	14.9	6.7	6.4	7.1	15.1	16.0	17.5	13.0	14.8	16.1
Rice (paddy)	20.5	21.4	22.6	9.4	8.1	8.8	40.0	32.2	29.0	17.6	16.9	16.3
Potatoes	–	77.1	92.8	51.9	51.8	56.3	81.6	99.3	93.8	117.6	106.2	123.0
	(.......PRODUCTION - 1,000,000 metric tons.......)											
Wheat	9.5	9.5	10.5	2.5	2.3	3.0	4.4	4.7	5.2	128.7	131.3	144.4
Rye	0.3	0.4	0.4	–	–	–	–	–	–	20.9	14.0	19.6
Barley	4.2	4.0	4.9	2.1	1.6	2.5	0.2	0.4	0.5	40.8	39.3	45.8
Oats	0.2	0.2	0.3	0.3	0.2	0.3	0.4	0.6	0.5	45.0	44.2	48.8
Maize	2.3	2.3	2.3	4.5	5.8	5.5	0.2	0.2	0.1	110.2	123.5	132.8
Rice (paddy)	1.6	2.2	2.2	1.6	1.9	2.3	0.1	0.1	0.1	151.2	145.2	152.6
Potatoes	0.3	0.7	0.9	0.5	0.7	0.8	0.5	0.7	0.6	159.1	133.3	157.0
Sugar beet / Sugar cane	0.2	0.4	0.4	1.0	1.1	1.4	1.8	1.4	1.9	26.3	27.0	34.0

APPENDIX II - LIVESTOCK NUMBERS

Region	Horses			Cattle			Pigs			Sheep		
	Pre-war	1945/46	1950/51	Pre-war	1945/46	1950/51	Pre-war	1945/46	1950/51	Pre-war	1945/46	1950/51
	(............................ Million head ....)											
Europe	20	15	16	102	86	100	78	40	77	119	94	112
N. & C. America	17	15	11	101	118	118	64	77	80	61	51	38
South America	18	17	18	106	123	136	30	34	35	97	115	123
Asia	13	10	10	227	219	219	83	70	83	127	107	125
Africa	3	3	3	77	78	90	3	3	4	106	100	112
Oceania	2	2	1	18	19	21	2	2	2	143	129	149
World (excl. USSR)	73	62	59	631	643	684	260	226	281	653	596	659

APPENDIX III - FOOD SUPPLIES PER HEAD (AT THE RETAIL LEVEL), PREWAR, RECENT AND TARGETS

Region Subregion and Country	Date	Cereals [1]	Starchy roots	Pulses	Sugar [2]	Fats [3]	Fruit [4]	Vegetables [4]	Meat [5]	Eggs	Fish [6]	Milk [7]
		(Kg. per head per year)										
FAR EAST												
South Asia												
Ceylon	1934–38	129	46	62	12	4	1	40	8	1	21	8
	1949/50	102	37	62	16	6	1	40	3	1	16	10
	1960 Target	117	43	62	12	6	50	75	3	1	16	10
India	[8] 1934–38	143	8	22	14	3	26	25	3	0.4	1	65
	1949/50	119	7	20	13	3	25	16	2	0.1	2	45
	1960 Target	129	8	29	13	5	50	70	3	0.1	3	50
Pakistan	1948/49	153	5	11	13	2	31	20	4	0.3	6	73
	1960 Target	159	6	13	13	2	60	80	5	1	7	87
East Asia												
China	1931–37	172	30	25	1	6	57		13	2	7	–
	1949/50	153	36	22	1	7	57		11	1	6	–
	1960 Target	153	38	32	1	6	40	90	16	3	8	1
Indochina	1934–38	144	18	5	7	2	48	71	14	3	6	7
	1949/50	125	17	4	4	1	...	...	5	2	5	3
	1960 Target	166	17	12	6	2	60	90	7	4	7	4
North-East Asia												
Japan	1934–38	168	63	8	14	2	16	79	4	3	35	5
	1949/50	157	66	2	3	1	12	66	2	1	27	4
	1960 Target	160	70	6	10	2	60	90	4	3	45	7
Pacific Is. and Malay Peninsula												
Indonesia	1934–38	130	131	7	6	6	16	44	5	2	9	1
	1949/50	120	118	8	2	6	...	...	4	1	6	2
	1960 Target	135	109	10	6	6	40	75	5	2	9	2
Philippines	1934–38	127	26	17	13	6	61	15	17	3	17	7
	1949/50	131	31	18	12	7	...	...	14	3	14	9
	1960 Target	143	30	34	13	7	80	70	17	4	21	14

For notes, see end of table.

APPENDIX III - FOOD SUPPLIES PER HEAD (AT THE RETAIL LEVEL), PREWAR, RECENT AND TARGETS *(continued)*

Region Subregion and Country	Date	Cereals [1]	Starchy roots	Pulses	Sugar [2]	Fats [3]	Fruit [4]	Vegetables [4]	Meat [5]	Eggs	Fish [6]	Milk [7]
		(Kg. per head per year)										
NEAR EAST												
Cyprus	1934–38	169	21	11	9	7	52	26	12	2	5	96
	1946/49	174	30	12	8	8	64	25	17	2	7	95
	1960 Target	157	30	13	9	10	75	65	20	3	9	120
Egypt	1934–38	182	5	21	10	3	36	33	11	2	4	55
	1946/49	174	10	13	11	3	36	58	10	1	4	55
	1960 Target	160	10	18	10	5	73	108	11	2	4	65
Iran	1934–38	162	2	9	5	1	75	43	12	3	4	93
	1946/49	144	1	8	6	1	66	42	10	2	5	84
	1960 Target	162	8	13	6	3	100	75	10	2	9	83
Iraq	1934–38	178	2	8	10	3	53	47	9	3	1	83
	1946/49	148	3	9	12	2	53	47	9	3	1	78
	1960 Target	170	3	13	11	3	80	75	11	4	3	85
Israel	[9] 1934–38	177	16	10	18	9	123	43	5	7	9	44
	1949/50	130	39	5	30	15	95	88	15	12	17	154
	1960 Target	130	50	11	20	14	102	90	15	20	17	155
Lebanon	1960 Target	137	21	24	11	9	175	93	20	4	3	109
Syria	[10] 1934–38	144	6	12	8	3	150	48	11	2	1	101
	[10] 1946/49	128	9	20	7	6	150	57	8	2	1	86
	1960 Target	153	20	24	8	7	148	75	7	3	1	78
Turkey	1934–38	191	3	8	5	3	98	53	22	3	5	151
	1946/49	177	9	7	6	3	120	65	19	2	4	141
	1960 Target	171	9	9	6	5	140	93	24	3	5	155

For notes, see end of table.

APPENDIX III - FOOD SUPPLIES PER HEAD (AT THE RETAIL LEVEL), PREWAR, RECENT AND TARGETS (*continued*)

Region Subregion and Country	Date	Cereals [1]	Starchy roots	Pulses	Sugar [2]	Fats [3]	Fruit [4]	Vege-tables [4]	Meat [5]	Eggs	Fish [6]	Milk [7]
		(............					Kg. per head per year					............)
AFRICA												
Northern Africa												
French N. Africa	1934–38	153	7	3	16	7	24	23	17	3	3	133
	1946/49	142	8	3	11	4	20	23	11	3	2	84
	1960 Target	156	10	6	14	5	60	40	13	3	4	99
Central & Tropical Africa												
Belgian Congo	1934–38	33	515	15	0.4	3	...	...	6	1	4	3
	1946/49	34	457	24	1	5	...	...	6	1	8	4
	1960 Target	49	451	33	1	6	40	60	7	2	8	5
French W. Africa	1934–38	148	70	16	1	...	...	...	10	1	1	18
	1946/48	125	133	22	1	...	...	...	12	1	1	24
	1960 Target	146	123	22	1	6	40	60	14	1	2	30
Kenya-Uganda	1934–38	58	491	26	5	3	25 (fruit and vegetables)		22	1	0.3	54
	1949/50	(58)	(491)	(26)	(5)	(3)	25 (fruit and vegetables)		(22)	(1)	0.2	(54)
	1960 Target	53	430	28	5	4	40	60	25	1	0.5	62
Tanganyika	1934–38	(87)	(306)	(18)	(2)	(1)	25 (fruit and vegetables)		(16)	(1)	0.4	(36)
	1949/50	87	306	18	2	1	25 (fruit and vegetables)		16	1	0.3	36
	1960 Target	99	303	24	2	2	40	60	19	1	0.7	44
Southern Africa												
Madagascar	1934–38	136	250	6	1	3	36 (fruit and vegetables)		29	2	4	138
	1946/49	115	208	6	4	2	35 (fruit and vegetables)		28	2	5	136
	1960 Target	127	197	7	4	3	40	60	28	2	6	152
Mauritius	1934–38	(135)	(28)	(6)	(42)	(7)	(34)	(33)	(7)	(2)	(8)	(36)
	1946/49	135	28	6	42	7	34	33	7	2	8	36
	1960 Target	132	25	9	39	7	39	65	8	3	11	43
Southern Rhodesia ...	1946/49	185	8	4	11	2	13	22	25	1	2	60
	1960 Target	176	9	6	11	3	40	60	27	2	3	65
Union of South Africa	1934–38	156	16	2	23	3	17	26	38	2	3	75
	1946/49	153	18	4	39	4	24	35	43	2	5	82
	1960 Target	145	18	4	38	6	40	60	43	3	7	82

For notes, see end of table.

APPENDIX III - FOOD SUPPLIES PER HEAD (AT THE RETAIL LEVEL), PREWAR, RECENT AND TARGETS (*continued*)

Region Subregion and Country	Date	Cereals [1]	Starchy roots	Pulses	Sugar [2]	Fats [3]	Fruit [4]	Vegetables [4]	Meat [5]	Eggs	Fish [6]	Milk [7]
		(............					Kg. per head per year					............)
LATIN AMERICA												
River Plate Countries												
Argentina	1935–39	106	66	3	27	10	47	25	107	7	6	162
	1946/49	125	87	2	35	17	58	39	114	7	5	164
	1960 Target	120	85	3	28	17	75	75	114	8	5	185
Paraguay	1935–39	58	247	30	29	9	(56)	...	118	3	0.4	129
	1946/49	69	249	27	38	7	...	...	94	3	0.3	104
	1960 Target	70	220	27	35	8	75	75	94	4	0.6	135
Uruguay	1935–39	85	40	3	24	13	29	10	107	7	3	166
	1946/49	100	34	2	28	12	40	14	107	8	2	183
	1960 Target	100	34	3	25	12	75	75	110	9	3	215
Rest of Latin America												
Brazil	1935–39	78	46	23	25	6	68	20	50	3	5	80
	1946/49	79	75	26	30	6	81	24	39	3	5	79
	1960 Target	82	75	30	28	7	85	75	45	4	5	90
Chile	1935–39	124	73	10	25	5	42	50	38	2	7	54
	1946/49	134	80	6	26	6	41	54	38	2	12	69
	1960 Target	140	80	10	26	7	75	90	43	3	15	82
Colombia	1935–39	57	87	7	40	3	132	10	26	4	0.3	94
	1946/49	72	98	8	62	3	105	12	29	4	1	145
	1960 Target	105	95	15	40	4	110	75	34	5	1	168
Cuba	1934–38	102	99	13	40	9	148	16	33	4	14	79
	1946/49	106	91	16	40	12	124	14	35	3	11	90
	1960 Target	102	88	18	40	12	135	75	42	5	12	105
Mexico	1934–38	109	5	9	18	5	43	...	25	3	1	86
	1946/49	123	7	10	26	6	58	24	23	2	2	71
	1960 Target	145	7	15	26	8	75	75	23	3	3	73
Peru	1934–38	98	109	16	14	4	42	14	24	3	1	39
	1946–49	103	122	7	22	4	43	14	23	3	5	36
	1960 Target	130	120	12	23	5	75	75	24	4	5	40
Venezuela	1935–39	93	78	16	33	3	112	4	21	2	7	98
	1946/49	95	58	16	43	4	96	10	26	3	17	87
	1960 Target	105	55	25	43	8	100	75	28	4	17	98

For notes, see end of table.

APPENDIX III - FOOD SUPPLIES PER HEAD (AT THE RETAIL LEVEL), PREWAR, RECENT AND TARGETS (*continued*)

Region Subregion and Country	Date	Cereals [1]	Starchy roots	Pulses	Sugar [2]	Fats [3]	Fruit [4]	Vegetables [4]	Meat [5]	Eggs	Fish [6]	Milk [7]
		(Kg. per head per year)										
EUROPE												
Western Europe												
Belgium-Luxembourg	1934–38	134	157	5	28	19	30	50	46	8	14	136
	1949/50	110	134	2	26	19	71	53	45	13	17	130
	1960 Target	115	133	3	26	19	78	82	47	15	18	143
France	1934–38	121	155	6	24	14	26	143	52	9	11	136
	1949/50	126	131	3	22	12	34	143	54	10	12	137
	1960 Target	128	130	4	22	12	50	149	54	11	13	156
Ireland	1934–38	131	195	1	38	14	20	53	55	16	6	149
	1949/50	126	189	1	35	21	19	58	56	14	6	193
	1960 Target	119	183	1	33	19	37	83	57	15	9	208
Netherlands	1934–38	98	142	3	34	22	42	64	40	6	8	203
	1949/50	94	176	2	36	23	53	62	29	5	12	219
	1960 Target	98	142	3	34	22	63	91	29	6	12	241
Switzerland	1934–38	110	91	2	38	15	86	62	56	9	2	307
	1949/50	116	98	2	39	16	78	72	40	9	3	344
	1960 Target	116	89	2	37	16	78	80	40	9	5	350
United Kingdom	1934–38	94	79	3	46	20	64	49	60	11	24	152
	1949/50	105	116	3	39	22	43	55	50	11	27	213
	1960 Target	104	91	3	38	20	50	70	54	12	27	235
Northern Europe												
Denmark	1934–38	94	123	—	51	27	33	62	75	8	18	195
	1949/50	107	140	1	31	19	38	64	65	9	18	221
	1960 Target	102	122	1	31	19	59	80	65	9	19	246
Finland	1934–38	128	181	3	28	13	18	30	33	3	14	276
	1949/50	145	184	3	25	12	3	28	27	4	15	271
	1960 Target	145	182	3	25	13	18	60	27	5	16	285

For notes, see end of table.

APPENDIX III - FOOD SUPPLIES PER HEAD (AT THE RETAIL LEVEL), PREWAR, RECENT AND TARGETS (*continued*)

Region Subregion and Country	Date	Cereals [1]	Starchy roots	Pulses	Sugar [2]	Fats [3]	Fruit [4]	Vegetables [4]	Meat [5]	Eggs	Fish [6]	Milk [7]
		(Kg. per head per year)										
Iceland	1934–38	120	70	1	44	15	4	18	51	4	129	313
	1948/49	105	77	1	35	19	12	15	80	5	118	390
	1960 Target	108	77	1	34	18	22	46	70	5	116	405
Norway	1934–38	119	131	2	30	25	33	19	38	7	42	207
	1949/50	123	127	3	23	25	19	20	31	7	55	301
	1960 Target	119	122	3	23	25	39	60	31	8	51	326
Sweden	1934–38	95	122	2	43	18	37	21	49	8	19	303
	1949/50	88	126	2	43	21	39	23	48	11	26	306
	1960 Target	86	118	2	40	21	54	60	46	11	26	322
Southern Europe												
Greece	1935–38	163	14	12	11	15	53	27	20	4	10	75
	1949/50	155	32	12	10	14	74	63	12	3	10	58
	1960 Target	159	31	15	9	14	77	85	13	5	12	65
Italy	1934–38	164	37	13	7	11	28	56	20	8	8	74
	1949/50	154	32	6	10	9	45	90	18	5	8	79
	1960 Target	162	36	12	11	13	53	90	20	6	9	90
Portugal	1934–38	103	76	10	10	13	60	80	23	2	30	51
	1948/49	120	90	10	12	10	41	72	19	2	31	43
	1960 Target	150	105	18	13	14	50	84	21	3	32	53
Spain	1931–35	146	109	15	12	15	50	100	28	5	25	73
	1948/49	126	91	14	9	12	35	115	23	4	18	68
	1960 Target	141	110	19	9	14	54	117	25	7	19	78
Eastern Europe												
Bulgaria	1934–38	222	9	11	3	8	37	86	22	4	1	103
	1949/50	210	7	9	10	7	50	88	18	5	1	87
	1960 Target	192	8	13	8	8	58	90	23	6	1	120

For notes, see end of table.

APPENDIX III - FOOD SUPPLIES PER HEAD (AT THE RETAIL LEVEL), PREWAR, RECENT AND TARGETS (*continued*)

Region Subregion and Country	Date	Cereals [1]	Starchy roots	Pulses	Sugar [2]	Fats [3]	Fruit [4]	Vegetables [4]	Meat [5]	Eggs	Fish [6]	Milk [7]
		(...					Kg. per head per year					...)
Czechoslovakia........	1934–38	130	160	4	24	14	42	41	33	8	5	124
	1948/49	139	145	3	23	10	46	65	34	6	7	105
	1960 Target	140	145	4	23	10	62	85	36	7	7	127
Hungary	1934–38	164	112	7	10	11	49	43	36	6	1	110
	1949/50	175	58	7	10	7	32	26	23	3	1	64
	1960 Target	168	100	10	9	11	62	82	30	4	1	90
Poland	1934–38	134	285	9	9	7	31	38	26	4	4	135
	1948/49	148	240	2	16	6	32	38	19	5	4	114
	1960 Target	149	240	8	14	7	47	70	21	5	6	137
Romania	1934–38	202	42	7	5	8	75	64	18	5	2	129
	1949/50	216	40	5	7	8	75	65	15	4	2	124
	1960 Target	170	39	7	8	8	77	78	18	6	3	138
Yugoslavia	1934–38	229	55	5	5	6	30	59	23	2	0.3	133
	1949/50	176	10	4	4	4	30	40	15	2	1	50
	1960 Target	181	20	6	5	6	39	71	20	3	1	80
Germany and Austria												
Austria	1934–38	138	96	2	24	18	43	58	49	7	2	199
	1949/50	137	106	1	21	15	38	60	29	4	3	125
	1960 Target	142	95	2	22	17	52	70	35	4	4	170
Eastern Germany	[11]1935–38	113	176	2	24	23	36	50	51	7	12	151
	1949/50	141	197	3	20	7	21	72	19	2	...	69
	1960 Target	148	193	6	19	10	47	100	25	3	...	81
Western Germany ...	[11]1935–38	113	176	2	24	23	36	50	51	7	12	151
	1949/50	120	194	3	23	15	31	43	23	4	19	109
	1960 Target	115	193	5	25	15	50	70	40	7	17	155

For notes, see end of table.

APPENDIX III - FOOD SUPPLIES PER HEAD (AT THE RETAIL LEVEL), PREWAR, RECENT AND TARGETS (*concluded*)

Region Subregion and Country	Date	Cereals [1]	Starchy roots	Pulses	Sugar [2]	Fats [3]	Fruit [4]	Vegetables [4]	Meat [5]	Eggs	Fish [6]	Milk [7]
		(Kg. per head per year)										
NORTH AMERICA												
Canada	1935–39	93	90	6	48	19	36	47	62	14	5	227
	1948/49	71	94	7	51	18	45	64	69	17	7	244
	1960 Target	71	80	7	47	17	55	86	73	17	7	266
United States of America	1935–39	90	64	7	49	20	86	98	64	16	5	249
	1948/49	78	49	7	47	19	90	113	74	21	5	289
	1960 Target	78	45	7	43	17	105	123	76	21	5	319
OCEANIA												
Australia............	1936/37-38/39	101	49	3	55	16	75	65	120	12	5	165
	1948/49	94	53	5	57	14	80	71	108	12	4	201
	1960 Target	90	47	4	51	14	92	101	112	15	4	225
New Zealand	1935–39	87	50	3	50	17	67	65	109	13	12	168
	1948/49	90	49	3	52	15	55	65	96	13	11	240
	1960 Target	85	47	3	45	14	67	86	97	13	13	255

1 Cereals: In terms of flour and meal.
2 Sugar: Including crude sugars consumed as such.
3 Fats: Including butter. Estimates expressed as far as possible as pure fats.
4 Fruit and Vegetables: Expressed, wherever possible, in fresh equivalent.
5 Meat: Expressed, wherever possible, in terms of dressed carcass weight.
6 Fish: Expressed, wherever possible, as fresh landed weight.
7 Milk: Excluding butter but including other milk products as fresh milk equivalent.
8 Includes Pakistan.
9 1934-38 data are for prewar Palestine.
10 Includes Lebanon.
11 1936 boundaries.

APPENDIX IV - CALORIE AND PROTEIN CONTENT OF NATIONAL AVERAGE FOOD SUPPLIES (AT RETAIL LEVEL)

Region Subregion and Country	Period	Calories per head per day	Total Protein	Animal Protein per head per day	Pulse Protein per head per day
			(................. Grams	....	.................)
FAR EAST					
South Asia					
Ceylon	1934–38	2140	48	9	9
	1949–50	1970	39	6	9
	1960 Target	2200	45	7	9
India	[1] 1934–38	1970	56	8	12
	1949–50	1700	44	6	10
	1960 Target	2000	53	7	15
Pakistan	1948–49	2020	52	11	7
	1960 Target	2230	59	13	8
East Asia					
China	1931–37	2230	72	7	19
	1949–50	2030	63	6	16
	1960 Target	2260	76	8	24
Indochina	1934–38	1850	44	7	2
	1949–50	1560	36	4	2
	1960 Target	2090	51	6	6
North-East Asia					
Japan	1934–38	2180	64	10	6
	1949–50	2100	53	8	1
	1960 Target	2210	60	14	4
Pacific Islands and Malay Peninsula					
Indonesia	1934–38	2040	46	5	6
	1949–50	1880	42	4	7
	1960 Target	2100	50	6	8
Philippines	1934–38	1920	45	11	2
	1949–50	1960	44	10	2
	1960 Target	2250	55	13	5
NEAR EAST					
Cyprus	1934–38	2340	65	11	6
	1946–49	2470	70	13	7
	1960 Target	2480	70	16	8
Egypt	1934–38	2410	76	12	13
	1946–49	2290	69	11	8
	1960 Target	2360	73	13	11
Iran	1934–38	2010	65	10	5
	1946–49	1820	58	9	5
	1960 Target	2150	69	11	8
Iraq	1934–38	2210	67	9	5
	1946–49	1930	60	8	6
	1960 Target	2280	72	10	9
Israel	[2] 1934–38	2550	73	11	6
	1949–50	2630	78	28	3
	1960 Target	2630	85	31	6
Lebanon	1960 Target	2540	76	15	15
Syria	[3] 1934–38	2040	63	10	7
	[3] 1946–49	2000	62	8	12
	1960 Target	2330	73	8	15
Turkey	1934–38	2600	92	27	5
	1946–49	2480	85	24	4
	1960 Target	2580	88	28	5

For notes, see end of table.

APPENDIX IV - CALORIE AND PROTEIN CONTENT OF NATIONAL AVERAGE FOOD SUPPLIES *(continued)*

Region Subregion and Country	Period	Calories per head per day	Total Protein	Animal Protein per head per day	Pulse Protein per head per day
			(.................	Grams	.................)
AFRICA					
Northern Africa					
French North Africa	1934–38 1946–49 1960 Target	2290 1920 2290	77 65 77	24 16 19	2 2 4
Central and Tropical Africa					
Belgian Congo	1934–38 1946–49 1960 Target	1910 1930 2190	37 42 52	4 5 6	9 14 19
French West Africa	1934–38 1946–48 1960 Target	2030 2070 2320	58 59 67	6 8 10	9 13 13
Kenya-Uganda	1934–38 1960 Target	2330 2250	58 60	14 16	15 16
Tanganyika........................	1934–38 1960 Target	1980 2230	53 62	10 12	11 14
Southern Africa					
Madagascar........................	1934–38 1946–49 1960 Target	2590 2250 2420	65 60 67	24 24 26	3 3 4
Mauritius.........................	1934–38 1946–49 1960 Target	2230 2230 2240	46 46 50	9 9 11	4 4 5
Southern Rhodesia	1934–38 1949 1960 Target	2300 2300 2340	68 68 70	16 16 18	2 2 3
Union of South Africa	1934–38 1946–49 1960 Target	2300 2520 2510	68 73 73	24 27 28	1 2 3
LATIN AMERICA					
River Plate Countries					
Argentina	1935–39 1946–49 1960 Target	2730 3190 3170	99 102 105	62 66 68	2 1 2
Paraguay..........................	1935–39 1946–49 1960 Target	2700 2670 2740	98 87 91	60 48 50	12 10 10
Uruguay	1935–39 1946–49 1960 Target	2380 2580 2720	90 94 102	61 62 67	1 1 2
Rest of Latin America					
Brazil	1935–39 1946–49 1960 Target	2150 2340 2470	68 64 71	32 26 27	13 14 16
Chile..............................	1935–39 1946–49 1960 Target	2240 2360 2600	69 72 83	21 23 28	6 4 6
Colombia	1935–39 1946–49 1960 Target	1860 2280 2590	47 56 74	20 26 30	4 5 9
Cuba	1934–38 1946–49 1960 Target	2630 2740 2820	64 68 74	25 26 31	8 10 11

For notes, see end of table.

APPENDIX IV - CALORIE AND PROTEIN CONTENT OF NATIONAL AVERAGE FOOD SUPPLIES (*continued*)

Region Subregion and Country	Period	Calories per head per day	Total Protein	Animal Protein per head per day	Pulse Protein per head per day
			(................. Grams	....	.................)
Mexico	1934–38	1800	53	18	5
	1946–49	2050	55	16	5
	1960 Target	2420	66	17	8
Peru	1934–38	1860	56	13	10
	1946–49	1920	52	14	4
	1960 Target	2350	66	15	7
Venezuela	1935–39	2040	57	19	10
	1946–49	2160	60	23	10
	1960 Target	2490	73	24	15
EUROPE					
Western Europe					
Belgium-Luxembourg	1934–38	3000	89	34	3
	1949–50	2770	81	36	2
	1960 Target	2880	86	38	2
France	1934–38	2880	88	37	4
	1949–50	2770	99	40	2
	1960 Target	2890	107	42	3
Ireland	1934–38	3390	99	48	1
	1949–50	3340	97	50	1
	1960 Target	3307	99	53	1
Netherlands	1934–38	3010	76	37	2
	1949–50	2960	83	40	2
	1960 Target	3030	90	45	2
Switzerland	1934–38	3110	95	54	1
	1949–50	3150	95	50	2
	1960 Target	3120	95	50	2
United Kingdom	1934–38	3100	82	45	2
	1949–50	3100	92	49	2
	1960 Target	3120	95	53	2
Northern Europe					
Denmark	1934–38	3390	88	54	–
	1949–50	3160	99	55	1
	1960 Target	3120	98	56	1
Finland	1934–38	3000	95	44	2
	1960 Target	3180	102	46	2
Iceland	1934–38	3160	111	74	1
	1960 Target	3240	123	83	1
Norway	1934–38	3160	86	46	2
	1949–50	3140	98	52	2
	1960 Target	3190	101	54	2
Sweden	1934–38	3080	89	54	1
	1949–50	3120	93	58	1
	1960 Target	3120	95	60	1
Southern Europe					
Greece	1935–38	2600	84	23	8
	1949–50	2510	80	18	9
	1960 Target	2634	87	21	12
Italy	1934–38	2510	82	20	9
	1949–50	2340	75	20	4
	1960 Target	2680	85	22	9
Portugal	1934–38	2110	66	22	7
	1960 Target	2730	84	21	12

For notes, see end of table.

APPENDIX IV - CALORIE AND PROTEIN CONTENT OF NATIONAL AVERAGE FOOD SUPPLIES (*concluded*)

Region Subregion and Country	Period	Calories per head per day	Total Protein	Animal Protein per head per day	Pulse Protein per head per day
			(.................	Grams	.................)
Southern Europe					
Spain	1931–35	2760	88	25	10
	1960 Target	2700	89	23	14
Eastern Europe					
Bulgaria	1960 Target	2800	87	21	8
Czechoslovakia	1960 Target	2810	88	36	2
Hungary...........................	1934–38	2770	82	25	4
	1960 Target	2730	82	21	6
Poland	1934–38	2710	79	23	6
	1960 Target	2780	82	25	5
Romania..........................	1960 Target	2680	82	22	5
Yugoslavia	1934–38	3020	95	22	4
	1960 Target	2440	76	16	4
Germany and Austria					
Austria	1934–38	2990	84	40	2
	1949–50	2620	75	27	1
	1960 Target	2860	85	35	1
Eastern Germany..................	1949–50	2460	72	19	2
	1960 Target	2710	80	22	4
Western Germany	[4] 1935–38	2960	83	40	2
	1949–50	2640	78	30	2
	1960 Target	2690	91	41	5
UNION OF SOCIALIST SOVIET REPUBLICS	1934–38	2830	88	17	–
	1949–50	3020	97	25	7
	1960 Target	3060	104	35	7
NORTH AMERICA					
Canada	1935–39	3070	85	48	4
	1948–49	3060	92	57	4
	1960 Target	3050	96	60	4
United States of America	1935–39	3150	89	50	4
	1948–49	3130	90	60	4
	1960 Target	3110	94	62	4
OCEANIA					
Australia	1936/37-1938/39	3300	103	67	1
	1948–49	3160	95	65	2
	1960 Target	3150	99	69	2
New Zealand......................	1935–39	3260	96	64	1
	1948–49	3250	96	65	1
	1960 Target	3180	99	68	1

1 Includes Pakistan.
2 1934-38 data are for prewar Palestine.
3 Includes Lebanon.
4 1935-38 data refer to Germany's 1936 boundaries.

Note: The prewar figures given here and in Appendix III as well, are not identical with those published in the first World Food Survey. Revised Food Balance Sheets for the prewar period have been computed and published (See FAO, *Food Balance Sheets*, 1949) for a considerable number of countries. Some countries have revised their prewar data on production, trade and consumption; better information on population has become available; and more accurate conversion factors have been computed. For a few countries the revised figures here presented differ considerably from those published in 1946.

APPENDIX V - CALORIE SUPPLIES (AT RETAIL LEVEL)

Region and Subregion	Population			Calories		
				Prewar	Recent	1960 Target
	1936	1949	1960	Calories per head per day	Calories per head per day	Calories per head per day
	(......... Millions ....)					
FAR EAST						
South Asia	385	445	505	1970	1770	2050
East Asia	508	532	559	2210	2000	2250
North-East Asia	94	114	128	2180	2100	2210
Pacific Islands and Malay Peninsula	91	103	118	2020	1900	2130
NEAR EAST	104	122	140	2320	2180	2370
AFRICA						
Northern Africa	18	23	28	2290	1920	2290
Central and Tropical Africa	85	99	112	2060	2080	2260
Southern Africa	24	31	37	2370	2420	2470
LATIN AMERICA						
River Plate Countries	17	20	23	2690	3090	3090
Rest of Latin America	106	138	164	2050	2250	2490
EUROPE						
Western Europe	114	120	125	3010	2950	3020
Northern Europe	17	19	27	3150	3120	3150
Southern Europe	82	92	100	2560	2350	2690
Eastern Europe	94	87	96	2840	2600	2700
Germany and Austria	65	76	81	2960	2600	2710
NORTH AMERICA and OCEANIA	148	174	198	3150	3130	3110
USSR	...	201	234	...	3010	3060

1960 TARGETS EXPRESSED AS INDICES IN RELATION TO PREWAR AND RECENT LEVELS

Calorie Supplies	Recent	1960 Target	Prewar	1960 Target
	(........ Prewar = 100 ....)		(........ Recent = 100 ....)	
World Population (excl. USSR)	112	125	89	111
(incl. USSR)	...	...	...	111
Total Calorie Supplies (excl. USSR)	106	128	94	121
(incl. USSR)	...	...	...	120
Average Calorie Supplies per head per day (excl. USSR)	94	102	106	105
(incl. USSR)	...	...	...	108

APPENDIX VI - GROSS FOOD SUPPLIES [1]

Region	Cereals [2]		Starchy Roots		Pulses [3]		Sugar [4]		Fats [5]	
	Million metric tons	Index (Recent = 100)	Million metric tons	Index (Recent = 100)	Million metric tons	Index (Recent = 100)	Million metric tons	Index (Recent = 100)	Million metric tons	Index (Recent = 100)
FAR EAST										
South Asia										
Prewar	77.73	96	4.03	107	13.66	88	6.18	97	1.16	75
Recent	81.25	100	3.77	100	15.59	100	6.39	100	1.55	100
1960 Target	101.72	125	5.32	141	21.12	135	7.37	115	2.41	156
East Asia										
Prewar	118.40	111	22.75	84	18.46	107	0.71	130	2.78	93
Recent	106.99	100	27.23	100	17.28	100	0.54	100	2.98	100
1960 Target	126.41	118	29.98	110	26.08	151	0.72	133	2.88	97
North-East Asia										
Prewar	20.12	96	7.22	63	2.04	339	1.37	467	0.24	95
Recent	20.94	100	11.40	100	0.60	100	0.29	100	0.26	100
1960 Target	26.63	127	14.29	125	2.06	342	1.37	468	0.34	134
Pacific Is. and Malay Peninsula										
Prewar	16.78	93	11.57	99	0.88	81	0.63	153	...	...
Recent	18.08	100	11.73	100	1.08	100	0.41	100	...	...
1960 Target	23.34	129	13.08	112	1.96	181	0.88	214	...	...
Total Region										
Prewar	233.03	103	45.57	84	35.03	101	8.89	116	...	...
Recent	227.26	100	54.14	100	34.56	100	7.64	100	...	...
1960 Target	278.09	122	62.67	116	51.22	148	10.34	135	...	...
NEAR EAST										
Prewar	30.05	94	0.74	50	1.67	98	0.74	70	0.32	74
Recent	32.13	100	1.49	100	1.71	100	1.05	100	0.43	100
1960 Target	41.51	129	2.47	166	2.75	161	1.22	116	0.81	188
AFRICA										
Northern Africa										
Prewar	4.38	94	0.19	79	0.09	110	0.30	119	0.12	130
Recent	4.65	100	0.24	100	0.08	100	0.25	100	0.10	100
1960 Target	6.06	130	0.38	158	0.21	257	0.37	149	0.12	125
Central and Tropical Africa										
Prewar	13.51	101	25.07	80	1.84	67	0.12	87	...	...
Recent	13.38	100	31.26	100	2.75	100	0.14	100	...	...
1960 Target	18.54	139	36.37	116	3.84	140	0.18	128	...	...
Southern Africa										
Prewar	5.24	74	2.54	123	0.18	63	0.34	52	0.11	84
Recent	7.09	100	2.06	100	0.29	100	0.64	100	0.13	100
1960 Target	8.93	126	2.29	111	0.56	194	0.80	125	0.22	170
Total Region										
Prewar	23.12	92	27.80	83	2.11	68	0.76	74	...	...
Recent	25.13	100	33.55	100	3.12	100	1.03	100	...	...
1960 Target	33.53	133	39.04	116	4.61	148	1.35	131	...	...
LATIN AMERICA										
River Plate Countries										
Prewar	6.45	82	2.14	70	0.15	107	0.48	69	...	...
Recent	7.84	100	3.06	100	0.14	100	0.70	100	...	...
1960 Target	8.33	106	3.64	119	0.20	142	0.68	98	...	...
Rest of Latin America										
Prewar	17.91	80	12.02	55	1.90	87	2.85	63	...	...
Recent	22.25	100	21.67	100	2.18	100	4.51	100	...	...
1960 Target	28.63	129	24.44	113	3.69	169	5.00	110	...	...
Total Region										
Prewar	24.35	81	14.16	57	2.05	88	3.34	64	...	...
Recent	30.09	100	24.73	100	2.32	100	5.25	100	...	...
1960 Target	36.96	123	28.08	114	3.89	168	5.69	108	....	...

For notes, see end of table.

APPENDIX VI - GROSS FOOD SUPPLIES [1]

Fruit [6]		Vegetables [7]		Meat [8]		Eggs		Fish [8,10]		Milk [9,10]		Region
Million metric tons	Index (Recent = 100)	Million metric tons	Index (Recent = 100)	Million metric tons	Index (Recent = 100)	Million metric tons	Index (Recent = 100)	Million metric tons	Index (Recent = 100)	Million metric tons	Index (Recent = 100)	
												FAR EAST
												South Asia
12.16	82	...	...	1.11	98	0.15	256	0.63	55	28.39	116	Prewar
14.81	100	9.96	100	1.14	100	0.06	100	1.14	100	24.48	100	Recent
34.63	234	46.33	465	1.44	127	0.11	193	1.90	166	31.42	128	1960 Target
												East Asia
...	...	...	...	5.98	120	0.72	109	3.17	111	0.16	167	Prewar
...	...	...	...	4.99	100	0.66	100	2.85	100	0.10	100	Recent
...	...	...	...	7.38	148	1.38	208	3.68	129	0.58	602	1960 Target
												North-East Asia
1.73	119	8.36	99	0.35	148	0.27	399	3.35	109	0.46	90	Prewar
1.46	100	8.48	100	0.24	100	0.07	100	3.06	100	0.52	100	Recent
9.14	627	14.34	169	0.55	231	0.36	524	6.16	201	1.07	206	1960 Target
												Pacific Is. and Malay Peninsula
...	...	...	...	0.62	97	0.16	91	0.96	119	0.16	48	Prewar
...	...	...	...	0.64	100	0.17	100	0.81	100	0.33	100	Recent
...	...	...	...	0.95	148	0.30	177	1.44	178	0.54	163	1960 Target
												Total Region
...	...	...	...	8.06	115	1.29	134	8.12	103	29.17	115	Prewar
...	...	...	...	7.00	100	0.96	100	7.86	100	25.43	100	Recent
...	...	...	...	10.32	147	2.15	224	13.18	168	33.61	132	1960 Target
												NEAR EAST
9.89	84	5.39	75	1.42	95	0.28	112	0.31	91	9.97	88	Prewar
11.77	100	7.22	100	1.50	100	0.25	100	0.34	100	11.36	100	Recent
19.74	168	15.44	214	2.15	143	0.47	188	0.51	150	14.92	131	1960 Target
												AFRICA
												Northern Africa
0.48	94	0.52	79	0.32	124	0.06	92	0.05	100	2.47	129	Prewar
0.51	100	0.66	100	0.26	100	0.06	100	0.05	100	1.92	100	Recent
1.66	325	1.10	167	0.36	138	0.10	151	0.10	206	2.67	139	1960 Target
												Central and Tropical Africa
...	...	...	...	0.88	74	0.09	90	0.13	63	1.80	75	Prewar
...	...	...	...	1.18	100	0.10	100	0.20	100	2.40	100	Recent
...	...	...	...	1.59	135	0.16	168	0.34	164	3.71	154	1960 Target
												Southern Africa
...	...	...	...	0.70	72	0.03	71	0.07	66	2.26	71	Prewar
...	...	...	...	0.98	100	0.05	100	0.11	100	3.18	100	Recent
...	...	...	...	1.27	130	0.09	188	0.19	178	4.22	133	1960 Target
												Total Region
...	...	...	...	1.90	79	0.18	86	0.25	69	6.54	87	Prewar
...	...	...	...	2.42	100	0.21	100	0.36	100	7.50	100	Recent
...	...	...	...	3.22	133	0.35	167	0.63	175	10.60	141	1960 Target
												LATIN AMERICA
												River Plate Countries
...	...	...	...	1.84	82	0.12	83	0.10	109	2.76	85	Prewar
...	...	...	...	2.24	100	0.14	100	0.09	100	3.26	100	Recent
1.72		1.72		2.57	114	0.18	126	0.11	121	4.24	130	1960 Target
												Rest of Latin America
7.75	75	2.16	70	4.03	91	0.30	81	0.42	61	8.57	77	Prewar
10.39	100	3.08	100	4.44	100	0.36	100	0.69	100	11.06	100	Recent
14.51	140	12.67	412	5.99	135	0.61	167	0.87	126	15.18	137	1960 Target
												Total Region
...	...	...	...	5.87	88	0.42	82	0.52	67	11.34	79	Prewar
...	...	...	...	6.68	100	0.51	100	0.78	100	14.32	100	Recent
16.23		14.39		8.56	128	0.79	155	0.98	126	19.42	136	1960 Target

APPENDIX VI - GROSS FOOD SUPPLIES[1] (*concluded*)

Region	Cereals[2]		Starchy Roots		Pulses[3]		Sugar[4]		Fats[5]	
	Million metric tons	Index (Recent = 100)	Million metric tons	Index (Recent = 100)	Million metric tons	Index (Recent = 100)	Million metric tons	Index (Recent = 100)	Million metric tons	Index (Recent = 100)
EUROPE										
Western Europe										
Prewar	42.25	105	29.90	88	0.76	151	3.98	105	2.91	96
Recent	40.06	100	33.94	100	0.50	100	3.79	100	3.04	100
1960 Target	44.00	110	39.21	116	0.60	120	3.85	101	3.15	103
Northern Europe										
Prewar	10.16	103	5.40	85	0.08	88	0.70	114	0.46	97
Recent	9.90	100	6.37	100	0.09	100	0.61	100	0.47	100
1960 Target	11.27	114	8.06	127	0.10	111	0.66	107	0.52	112
Southern Europe										
Prewar	24.69	114	7.77	106	1.78	160	0.73	83	1.34	118
Recent	21.75	100	7.35	100	1.11	100	0.87	100	1.14	100
1960 Target	28.82	133	9.44	128	1.78	160	1.03	118	1.57	138
Eastern Europe										
Prewar	40.31	111	52.11	143	1.00	176	0.90	90	0.97	138
Recent	36.18	100	36.47	100	0.57	100	1.00	100	0.70	100
1960 Target	41.24	114	44.39	122	1.14	200	1.08	108	0.98	139
Austria and Germany										
Prewar	27.60	124	51.33	156	0.65	159	1.78	105	2.23	218
Recent	22.19	100	32.92	100	0.41	100	1.71	100	1.02	100
1960 Target	24.95	112	40.32	122	0.56	138	1.88	110	1.20	117
Total Region										
Prewar	145.00	111	146.50	125	4.27	159	8.10	101	7.91	124
Recent	130.08	100	117.04	100	2.68	100	7.99	100	6.38	100
1960 Target	150.29	116	141.41	121	4.18	156	8.49	106	7.42	116
NORTH AMERICA AND OCEANIA										
Prewar	102.56	78	14.14	92	1.62	82	8.07	97	3.03	94
Recent	132.22	100	15.31	100	1.98	100	8.32	100	3.23	100
1960 Target	153.00	116	13.32	87	2.24	113	9.04	109	3.36	104
USSR										
Recent	107.0	100	86.0	100	2.9	100	3.00	100	1.90	100
1960 Target	122.0	114	85.0	99	3.6	124	3.32	111	2.50	132
World Total (excl. USSR)										
Prewar	558	97	249	101	47	101	30	96	...	...
Recent	577	100	246	100	46	100	31	100	...	...
1960 Target	693	120	287	117	69	149	36	116	...	...
World Total (incl. USSR)										
Recent	684	100	332	100	49	100	34	100	...	...
1960 Target	815	119	372	112	72	147	39	115	...	...

1 Gross supplies: The tonnage figures shown for gross supplies for the 1960 targets have been calculated, except where stated below, by the method explained on page 21 of the text. The corresponding figures for prewar and recent postwar years have been extracted from actual food balance sheets. Because of the lack of precision of much of the data, especially on crop utilization, as well as the numerous assumptions involved in the computation of the 1960 targets, the tonnage figures shown should not be considered more than a rough indication of the orders of magnitude of gross supplies. The chief significance of the above table lies in the percentage changes for the three periods shown by the index of gross supplies.

2 Cereals: Expressed in terms of the actual weight of cereals.

3 Pulses: Including nuts, except in Europe. Also including soybeans in the Far East.

4 Sugar: Including crude sugars produced in Latin America and elsewhere.

APPENDIX VI - GROSS FOOD SUPPLIES [1] (*concluded*)

Fruit [6]		Vegetables [6]		Meat [7]		Eggs		Fish [8,10]		Milk [9,10]		
Million metric tons	Index (Recent = 100)	Million metric tons	Index (Recent = 100)	Million metric tons	Index (Recent = 100)	Million metric tons	Index (Recent = 100)	Million metric tons	Index (Recent = 100)	Million metric tons	Index (Recent = 100)	Region
												EUROPE
												Western Europe
5.58	90	12.11	95	6.17	106	1.10	90	1.79	84	17.51	79	Prewar
6.21	100	12.72	100	5.82	100	1.23	100	2.13	100	22.07	100	Recent
8.12	131	15.09	119	6.34	109	1.37	111	2.33	109	25.60	116	1960 Target
												Northern Europe
0.56	97	0.67	92	0.82	99	0.11	73	0.38	74	4.28	82	Prewar
0.58	100	0.73	100	0.83	100	0.15	100	0.51	100	5.20	100	Recent
1.10	190	1.60	220	0.90	109	0.18	118	0.57	112	6.19	119	1960 Target
												Southern Europe
3.87	84	6.16	68	1.81	104	0.49	126	1.23	105	5.80	91	Prewar
4.60	100	9.02	100	1.74	100	0.39	10	1.17	100	6.34	100	Recent
6.53	142	11.54	128	2.11	121	0.58	150	1.42	121	7.96	125	1960 Target
												Eastern Europe
5.32	108	5.83	115	2.49	147	0.44	131	0.26	112	12.15	152	Prewar
4.91	100	5.07	100	1.70	100	0.34	100	0.24	100	8.01	100	Recent
7.27	148	9.00	178	2.22	131	0.46	138	0.34	144	11.26	140	1960 Target
												Austria and Germany
3.45	131	4.48	94	3.78	219	0.54	203	0.82	62	11.53	151	Prewar
2.64	100	4.76	100	1.72	100	0.26	100	1.32	100	7.65	100	Recent
5.01	190	7.51	158	2.94	171	0.48	181	1.32	100	11.26	147	1960 Target
												Total Region
18.78	99	29.25	91	15.08	128	2.68	113	4.49	84	51.26	104	Prewar
18.93	100	32.30	100	11.81	100	2.37	100	5.37	100	49.28	100	Recent
28.04	148	44.47	139	14.52	123	3.07	130	5.98	111	62.27	126	1960 Target
												NORTH AMERICA AND OCEANIA
13.17	86	16.09	81	11.28	79	2.52	67	0.79	88	65.82	96	Prewar
15.29	100	19.85	100	14.32	100	3.77	100	0.90	100	68.52	100	Recent
21.24	139	26.55	134	17.17	120	4.23	112	1.06	118	85.97	125	1960 Target
												USSR
16.3		100		4.30	100	0.20	100	1.4	100	31.0	100	Recent
38.0		233		6.50	151	0.40	200	2.7	193	49.0	158	1960 Target
												World Total (excl. USSR)
...	...	...	...	44	100	7.4	91	14	93	174	99	Prewar
...	...	...	...	44	100	8.1	100	16	100	176	100	Recent
...	...	...	...	56	128	11.1	137	22	143	227	129	1960 Target
												World Total (incl. USSR)
...	...	...	...	48	100	8.3	100	17	100	207	100	Recent
...	...	...	...	62	130	11.5	139	25	147	276	133	1960 Target

5 Fats: Net supplies for human consumption only in Latin America and in a number of other countries where a gross supply figure could not be computed.

6 Fruit and Vegetables: Expressed as far as possible in terms of fresh equivalents. The figures represent net supplies for countries where consumption data only are available.

7 Meat: Expressed in terms of carcass weight including offal, poultry, and game.

8 Fish: Expressed in terms of landed weight as far as possible.

9 Milk: Including milk products other than butter in terms of fresh milk.

10 Net supplies for human consumption only.

APPENDIX VII - REGIONAL AND SUBREGIONAL GROUPS

(* *indicates countries for which food supply information is available*)

EUROPE

Western Europe

Belgium-Luxembourg *
France *
Ireland *
Netherlands *
Switzerland *
United Kingdom *

Channel Islands
Isle of Man
Liechtenstein
Monaco

Northern Europe

Denmark *
Finland *
Iceland *
Norway *
Sweden *

The Faeroes
Svalbard

Austria and Germany

Austria *
Eastern Germany *
Western Germany *

Southern Europe

Greece *
Italy *
Portugal *
Spain *

Andorra
Gibraltar
Malta and Gozo
San Marino
Trieste
Vatican City

Eastern Europe

Bulgaria *
Czechoslovakia *
Hungary *
Poland *
Romania *

Albania
Yugoslavia

UNION OF SOCIALIST SOVIET REPUBLICS *

NORTH AMERICA AND OCEANIA

Canada *
United States of America *
Australia *
New Zealand *

NORTH AMERICA AND OCEANIA (*cont.*)

Alaska
Bermuda
Greenland
St. Pierre and Miquelon

FAR EAST

South Asia

Ceylon *
India *
Pakistan *

Bhutan
French India
Macao
Nepal
Portuguese India

East Asia

China *
Indochina *

Burma
Hong Kong
Mongolian People's Republic
Thailand

North-East Asia

Japan *

Korea
Ryukyu Islands

Pacific Islands and Malay Peninsula

Indonesia *
Philippines *

American Samoa
British Borneo
Fiji Islands
French Oceania
Gilbert and Ellice Islands
Guam
Hawaii
Malaya, Federation of
Maldive Islands
Nauru
New Caledonia
New Guinea
New Hebrides
Portuguese Timor
Singapore *
Solomon Islands
Tonga
U. S. Pacific Islands
West Samoa

NEAR EAST

Cyprus *
Egypt *
Iran *
Iraq *
Israel *
Lebanon *
Syria *
Turkey *

Aden
Afghanistan
Anglo-Egyptian Sudan
Bahrein
Eritrea
Ethiopia
Jordan
Kuwait
Muscat and Oman
Qatar
Saudi Arabia
Trucial Oman
Yemen

AFRICA

Northern Africa

French North Africa *

Canary Islands
Libya
Madeira Islands
Spanish Morocco
Tangier

Central and Tropical Africa

Belgian Congo *
French West Africa *
Kenya-Uganda *
Tanganyika *

Angola
British Cameroons
British Somaliland
British Togoland
French Cameroons
French Equatorial Africa
French Somaliland
French Togoland
Gambia
Gold Coast
Liberia
Nigeria
Portuguese Guinea
Ruanda Urundi
São Tomé
Sierra Leone
Somalia
Spanish Guinca

Central and Tropical Africa (cont.)

Spanish West Africa
Zanzibar

Southern Africa

Madagascar *
Mauritius *
Southern Rhodesia *
Union of South Africa *

Basutoland
Bechuanaland
Mozambique
Northern Rhodesia
Nyasaland
Réunion
Seychelles
Swaziland

LATIN AMERICA

River Plate Countries

Argentina *
Paraguay *
Uruguay *

Rest of Latin America

Brazil *
Chile *
Colombia *
Cuba *
Mexico *
Peru *
Venezuela *

Bolivia
British West Indies
British Guiana

LATIN AMERICA (*cont.*)

British Honduras
Costa Rica
Dominican Republic
Ecuador
El Salvador
French Guiana
Guadeloupe
Guatemala
Haiti
Honduras
Martinique
Netherlands West Indies
Nicaragua
Panama (incl. Canal Zone)
Puerto Rico
Surinam
Virgin Islands

SQUARCI - ROMA - VIA LABICANA 92 - TEL. 71.006

Freedom from Hunger Campaign

BASIC STUDY
Nº 11

THIRD WORLD FOOD SURVEY

FAO
FIAT PANIS
FREEDOM FROM HUNGER

FOOD AND AGRICULTURE ORGANIZATION OF THE UNITED NATIONS • ROME • ITALY

FREEDOM FROM HUNGER CAMPAIGN

Basic Studies Series

A series of Basic Studies supporting the Freedom from Hunger Campaign is being published by the Food and Agriculture Organization and other organizations of the United Nations.

Sixteen such studies are contemplated in the series; nine will be issued by the Food and Agriculture Organization (FAO), three by the United Nations, and one each by the United Nations Educational, Scientific and Cultural Organization (UNESCO), the International Labour Organisation (ILO), the World Health Organization (WHO) and the World Meteorological Organization (WMO).

The subjects cover a wide range and include, for example, the possibilities of increasing world food production, the part marketing can play in increasing productivity, education and training in nutrition, animal disease and human health, economic development through food, population and food supplies, education in relation to agriculture and economic development, hunger and social policy, malnutrition and disease, weather and food.

This volume, *Third world food survey*, issued by the Food and Agriculture Organization, is No. 11 in the series.

The following titles have already been issued:

No. 1. *Weather and food*, WMO, Geneva.

No. 2. *Development through food - A strategy for surplus utilization*, FAO, Rome.

No. 3. *Animal disease and human health*, FAO, Rome.

No. 4. *Marketing - Its role in increasing productivity*, FAO, Rome.

No. 5. *Nutrition and working efficiency*, FAO, Rome.

No. 6. *Education and training in nutrition*, FAO, Rome.

No. 7. *Population and food supplies*, United Nations, New York.

No. 8. *Aspects of economic development*, United Nations, New York.

No. 9. *Increasing food production through education, research, and extension*, FAO, Rome.

No. 10. *Possibilities of increasing world food production*, FAO, Rome.

THIRD WORLD FOOD SURVEY

SQUARCI - ROME

FFHC Basic Study *No. 11*

THIRD WORLD FOOD SURVEY

FOOD AND AGRICULTURE ORGANIZATION
OF THE UNITED NATIONS Rome 1963

Printed in Italy

CONTENTS

PREFACE

Perhaps one of the most arresting conclusions to be drawn from this publication, which is the third of a series of world food surveys, is that while the world food consumption level has improved over the last decade, up to half the world's population is still hungry or malnourished or both. The survey shows that most of the improvement occurred in the developed areas while the improvement made in the less developed areas was hardly enough to regain the unsatisfactory prewar level. It thus confirms once again that the gap between the developed and the less developed regions tends to increase rather than to decrease.

The survey presents a comprehensive picture of the present and past world food situation. It is based on food balance sheet data for over 80 countries covering some 95 percent of the world's population. It also draws upon the food consumption and dietary surveys conducted in various parts of the world and introduces new statistical techniques in the study of food supplies and needs.

FAO had only been established a few months when it produced the first *World food survey.* This was a pioneer effort to appraise the world food situation, based on data from 70 countries representing about 90 percent of the world population at that time. While there were gaps in the statistical information then available and much of the material used was in the nature of intelligent guesswork, it did show that food supplies were inadequate over large areas of the world.

A somewhat more reliable and more detailed picture of the food situation was presented by FAO's *Second world food survey* which was issued in 1952. It was based on data for food supplies in the postwar period and revised data for the prewar period. Further, more realistic standards for calorie requirements had been established, taking into account such factors as age, sex, body weight, and average activity of populations living in differing environmental temperatures. The analysis showed that in areas containing some 60 percent of the world population food supplies at the retail level were not sufficient to provide even 2,200 Calories per day.

In this latest report, the *Third world food survey*, the point that emerges time and again is that, by any measurement, the diets of the mass of people in the less developed countries and regions remains appallingly low. The people in the less developed countries are found to have a Calorie intake of 2,150 per day compared with 3,050 for the people in the developed countries. The differences in quality of diet are even more striking. For instance, the intake of animal protein in these countries and regions is only one-fifth of that in the more developed areas. The intakes in the less developed countries are considerably short of the requirements and in fact it is estimated that at least 20 percent of the population in these areas is undernourished and 60 percent malnourished. Nobody, then, should be astonished by the conclusion reached by the survey that 10-15 percent of the people in the world are undernourished " and up to half suffer from hunger or malnutrition or both. " These facts, together with the rapidly expanding population, present a very serious challenge to mankind. Even looking only to the needs of the development decade (i.e., by 1975), the world food supplies will have to be increased by more than 35 percent merely " to sustain the world's population at its present unsatisfactory level of diet. " If, as we should, we are to achieve a reasonable improvement in the level of nutrition, then world food supplies will have to be increased by more than 50 percent. The figures for the less developed countries will have to be much higher. It is estimated that by 1975 total food supplies in these countries will need to be increased by about 80 percent and that the production of animal products, in particular, will need to be raised by more than 120 percent.

Even more demanding are the long-term requirements. By the year 2000 it is expected that the world population will be at least double the present figure, while the population in the less developed countries will have increased by 150 percent. If this proves true, then these less developed countries will need to increase their total food supplies to four times the present volume and their supplies of animal food products to about sixfold. These figures give some indication of the magnitude of the task confronting us, and indicate how timely and urgent is the Freedom from Hunger Campaign.

The Campaign, which is supported by all the member nations of FAO, has already contributed widely toward knowledge and understanding of the world hunger problem. It has greatly helped to bring the problem to the forefront in international discussions which have led to many valuable developments. Among these is the setting up of the World Food Program, which is being operated jointly by the United Nations and FAO. The campaign can also be expected to do much to help promote the UN Development Decade. Indeed, the Freedom from Hunger Campaign will be FAO's contribution to the Development Decade.

There is one overriding consideration to be kept in mind. It is this: the Freedom from Hunger Campaign is an effort which must be continued for many years. Even with a climate of opinion which would enable the hunger problem to be met on the national and international scale required, the magnitude of the task, as shown by this survey, is such that decades and generations of work and effort lie ahead.

B. R. Sen
Director-General

1. BACKGROUND AND CONCLUSIONS

Like the first and second world food surveys, the third world food survey is concerned with these fundamental questions: What is the food consumption of the populations of the different countries? How does it compare with their needs? Where are the most serious shortages? What kinds of food and what quantities of each are needed to achieve improvement in nutrition throughout the world? What are the technical, economic, and social factors underlying this aim?

The first *World food survey* was published in 1946, only a few months after FAO had been established, at a time when statistical services of many countries were still disrupted and the prewar period was the latest for which information on a world-wide basis was available. Admittedly, there were large gaps in the statistical information relating to this period. For many countries the prewar figures of food supplies and sometimes of population were in fact mostly guesses based partly on a critical examination of whatever material was obtainable in scattered sources and partly on the subjective judgment of informed experts. Nevertheless, the first world food survey served a useful purpose. A pioneer attempt to appraise the world food situation, it gave a more studied and detailed picture of prewar conditions than had until then been available. It disclosed the main gaps between consumption and nutritional requirements and called attention to the possibilities for closing these gaps. On the basis of data for 70 countries whose people made up some 90% of the world's population, it was found that

" In areas containing over half the world's population, food supplies at the retail level (not actual intake) were sufficient to furnish an average of less than 2,250 Calories * per caput daily. Food sup-

* Calorie = kilocalorie.

plies furnishing an average of more than 2,750 Calories per caput daily were available in areas containing somewhat less than a third of the world's population. The remaining areas, containing about one sixth of the world's population, had food supplies that were between these high and low levels."

The *Second world food survey* was published in 1952. It was essentially concerned with the same basic questions as the first, viewed in the light of changes that had occurred and the greater knowledge available. By that time current statistics from many countries were more readily obtained than in 1946 and the food balance sheet method of estimating food supplies for human consumption had been considerably improved. Data on food supplies for the postwar period and also revised data for the prewar period could be prepared for 52 countries covering some 80% of the world's population. Moreover, a distinct advance had taken place in establishing standards for calorie requirements for individual countries based on age and sex composition of the population, the reference body weight, average physical activity and the mean environmental temperature. While in the first world food survey a uniform figure of 2,600 Calories per person per day was recommended for the determination of the adequacy of calorie levels, the analysis in the second world food survey was based on this newly devised calorie-requirement scale, recommended by a group of international experts convened by FAO.

The results of the survey showed that in areas covering 59% of the world's population, food supplies at the retail level were not sufficient to furnish an average of 2,200 Calories daily. Food supplies furnishing an average of over 2,700 Calories per caput daily were available in areas containing 28% of the world's population. The remaining areas, containing 13% of the world's population, had food supplies that were between these high and low levels, the corresponding prewar figures being 39, 30, and 31%. Thus, five years after the second world war when most countries were still engaged in repairing war ravages, the average calorie supply per person over large areas of the world was still lower than before the war. This was particularly so in the less developed areas, contributing to further widening of the already large gaps between them and the more developed areas. Nevertheless, the gaps between the actual and the desired appeared much less in 1952 than in 1946 because the new calorie-requirement scale for individual countries, and particularly for those with low intake levels, implied lower calorie-requirement figures than the global figure of 2,600 Calories used in the first world food survey. As regards the qualitative aspect of the diet, it was found that in the areas with low calorie intake the animal protein content of the diet fell well below the already low prewar levels. Evidence of poor quality diet was also afforded by the proportion of calories derived from cereals, starchy roots, and sugar, which appeared unduly high and exceeded two thirds for most of the low-calorie countries. The high proportions of prewar years were maintained, and in some regions like the Far East and the Near East they tended to increase, indicating again that the position compared with prewar noticeably worsened.

The *Third world food survey*, like the second, compares the present situation with the past and takes account of the trends and changes that have taken place since prewar. It thus covers three periods, prewar, postwar, and recent; and gives an outlook of the needs and possibilities for the future. In general, prewar data refer to the period 1934-38, postwar data refer to the period 1948-52, and the recent to 1957-59.[1] Deviations from these time references were dictated by the availability of information. Since the preparation of the second world food survey, more and better data have become available, though they still vary a good deal between the different parts of the world with regard to coverage, concepts, and reliability.

For example, food balance sheets, admittedly provisional, have been prepared for the U.S.S.R. and Mainland China and many other countries for which information had previously been scanty or for which estimates were based on mere intelligent guesses. This survey has been based on food balance sheet data for over 80 countries covering some 95% of the world's population. Food consumption and dietary surveys conducted in various parts of the world have provided a wealth of supplementary information. Current statistics of food production and utilization have improved in quality as a result of

---

[1] Although agricultural data in general refer to split years, the usual stroke notation (e.g., 1957/58-1959/60) has been omitted in text and text tables for reasons of simplification.

the introduction of objective methods of collecting data in a number of countries. Better methods of analysis have also become available. Thus, in the first world food survey the determination of the quantitative adequacy of diets had to be made against a uniform standard of 2,600 Calories per caput per day and in the second survey such analysis was confined to countrywise comparison of the levels of intake and requirements based on the FAO calorie-requirement scale. Recently more refined techniques based on the concept of variation in requirements among individuals and the revised FAO requirement scale have become available for estimating the incidence of undernutrition from data collected in household food consumption surveys. Much the same conditions are true regarding the estimation of the incidence of malnutrition.

The broad conclusions of the third world food survey are as follows:

1. The world food supply available per caput, though higher than in the postwar years, is only slightly above the prewar level. The progress over the last decade has mainly taken place in the developed areas with the result that the gap between the developed and the less developed areas has tended to increase rather than to decrease.

2. The calorie content of the diet has generally regained the prewar level in both the developed and the less developed areas. Nevertheless, the current calorie supply per caput in the less developed areas falls short of the corresponding requirement and it is estimated that at least 20% of the population in these areas is undernourished.

3. The nutritional quality of the diet has shown a distinct though small improvement over the prewar level. However, this improvement has again mainly taken place in the developed areas, while in the less developed areas the quality of diets has barely regained the unsatisfactory prewar level. Reports of dietary and clinical surveys show that nutritional deficiency diseases are still common in large parts of the world. Retarded growth of children, poor physique and health of adults, low resistance to disease, particularly in children below 5 years, and low working efficiency together with high mortality rates among young children and low expectations of life are an indication of widespread malnutrition in the less developed

areas. This is not surprising, since in these areas the level of animal protein intake is only one fifth of that in the more developed areas.

4. It is generally agreed that if more than about 80% of the calories in a diet are derived from cereals, starchy roots, and sugar, there is a risk that the nutritional quality of diet is inadequate. If this percentage is less than about 80, the diet is likely to be adequate. In well-fed countries like the United Kingdom and France hardly any households derive more than 80% of their calories from cereals, starchy roots, and sugar, whereas some 60% of the households in the less developed countries have a proportion exceeding 80%. In other words, it appears probable that some 60% of the households in the less developed areas live on diets which are inadequate in nutritional quality.

5. Presenting the results of undernutrition and malnutrition for the world as a whole, the survey concludes that 10 to 15% of its people are undernourished and up to a half suffer from hunger or malnutrition or both.

6. If the world population were to grow according to the United Nations projections under the "medium" assumption (and recent indications are that the population growth is likely to be larger), then by 1975 world food supplies would need to be increased by over 35% merely to sustain the world's population at its present unsatisfactory level of diet. If, in addition, a reasonable improvement in the level of nutrition is to be brought about, world food supplies would have to be increased by over 50%, and in particular, food supplies of animal products would have to be increased by some 60%. In the less developed areas where the population increase will be faster, the corresponding figures are much higher. In these areas total food supplies will have to be increased by some 80% and those of animal foods by over 120%.

7. In order to achieve these targets by 1975 the less developed countries would need to aim at an annual rate of increase in per caput food supplies approaching 2%. Since the income elasticity of the demand for food in these countries may be expected to be of the order of .7, this in effect implies a rate of growth in per caput income approaching 3%. Should the population grow at a rate of

about 2.0% suggested by the United Nations projections, the aim should be to increase the aggregate national income by some 5% per annum, which is compatible with the target for the decade 1960/70 in the plans for activating economic development through the joint efforts of the United Nations and its Specialized Agencies.

8. The targets for 1975 are only a first step in improving the level of food consumption and nutrition. The survey also looks further ahead and concludes that should the population grow according to the United Nations " medium " projection, the world's total food supply would have to be trebled by the year 2000 in order to provide a reasonably adequate level of nutrition. For the less developed areas total food supplies would need to be quadrupled and the supplies of animal products should be raised to six times the present volume.

It is hoped that the facts presented in this survey will be found of assistance in formulating plans and programs for more intense and comprehensive action to combat hunger and malnutrition in the world.

2. DEVELOPMENTS IN THE WORLD FOOD SITUATION

Population growth

The world's population was estimated to grow at about 1% per annum when the first world food survey was published in 1946. There was no appreciable increase in the rate of growth in population even when the second world food survey was published in 1952. The successful control of epidemics and diseases in large parts of the world since the beginning of the last decade has, among other causes, led to a marked lowering of death rates and as a result to an accelerated growth in the world's population, as indicated in Table 1.

TABLE 1. - POPULATION AND POPULATION GROWTH, BY GROUPS OF REGIONS

Area	1938	1950	1960	Annual percentage rate of growth (compound)	
				1938-50	1950-60
	....	*Millions*	....		
Less developed regions [1]	1 478	1 733	2 161	1.3	2.2
Developed regions [2]	717	751	852	0.4	1.3
World	2 195	2 484	3 014	1.0	1.9

[1] Included here and in the following tables: Far East, China, Mainland; Near East; Africa; Latin America.
[2] Included here and in the following tables: Europe, U.S.S.R.; North America; Oceania.

As against a growth rate of 1% per annum during the period 1938-50, the population grew at nearly *twice* that rate during the last decade. In particular, Table 1 shows that the rate of growth in the less developed regions shot up from an estimated 1.3% per annum during 1938-50 to 2.2% during 1950-60. Although the developed regions also show a comparable increase in growth rate, the current estimated growth rate in these areas is still not higher than that in the less developed regions a decade ago. Of the total increase of 800 millions between 1938 and 1960 in the world's population, the less developed regions account for over 650 millions, with the Far East alone responsible for nearly 500 millions (Appendix 1). The figures point to the increasing share of the less developed areas in the world's population. Whereas in 1938 the less developed areas accounted for some 67% of the world's total population, the estimated share today is some 72%. This unprecedented rate of population growth in less than a decade is the most striking feature of the developments since the publication of the second world food survey. As we shall see later it has tremendous implications for future food needs if we are to ensure a reasonably good level of nutrition to the peoples of the world.

Food production

Appendix 2 shows the trends in food production by regions since prewar years. Table 2 summarizes these trends for the less developed and the developed regions.

TABLE 2. - INDEX NUMBERS OF FOOD PRODUCTION, BY GROUPS OF REGIONS [1]
(PREWAR = 100)

Area	Average 1948-52	Average 1957-61
Less developed regions (excl. China, Mainland) . .	115	150
Less developed regions (incl. China, Mainland) . .	107	145
Developed regions	117	152
World (excl. China, Mainland)	116	152
World (incl. China, Mainland)	112	149

[1] These index numbers have been calculated by applying regional weights, based on 1952-56 farm price relationships, to the production figures, which are adjusted to allow for quantities used for feed and seed. The same holds for Tables 5-7. The index numbers in Tables 2 and 5 are also shown excluding Mainland China for reasons of consistency with other FAO publications.

Since prewar years, world food production has increased by some 50%. The percentage increase is about the same in both the less developed and the developed regions. In the developed regions this increase is mainly due to a rise in yields; in fact, area under many

TABLE 3. - INDEX NUMBERS OF AREA, YIELD, AND PRODUCTION OF MAJOR CROPS BY GROUPS OF REGIONS (PREWAR = 100)

Item	Less developed regions		Developed regions		World	
	Average 1948-52	Average 1957-60	Average 1948-52	Average 1957-60	Average 1948-52	Average 1957-60
Wheat						
Area	103	128	99	117	101	121
Yield	86	102	109	129	101	120
Production	89	129	108	152	101	145
Rice, paddy						
Area	119	139	160	140	119	139
Yield	91	109	89	120	91	110
Production	108	151	143	168	109	151
Other cereals						
Area	122	135	94	93	106	111
Yield	85	105	113	146	100	124
Production	104	141	106	135	105	137
Starchy roots						
Area	168	263	94	95	115	143
Yield	100	110	106	113	98	100
Production	169	289	99	107	113	143
Total pulses						
Area	117	141	102	93	130	130
Yield	94	94	90	103	82	95
Production	110	133	91	96	106	125

cereals, starchy roots, and pulses has declined, while in the less developed regions the increase in production is more often achieved along traditional lines by bringing more land into cultivation (Table 3).

The corresponding developments for livestock products are illustrated in Table 4 by the example of cattle.

TABLE 4. - INDEX NUMBERS OF CATTLE NUMBERS, CATTLE PRODUCTS OUTPUT[1] AND YIELD, BY GROUPS OF REGIONS (AVERAGE 1948-52 = 100)

Area	Cattle numbers	Yield	Output
	Average 1958-60	Average 1958-60	Average 1958-60
Less developed regions	114	106	122
Developed regions..........	117	114	133
World	115	113	130

[1] Meat and milk in terms of milk equivalent taking 1 unit of meat as equal to 10 units of milk.

The increases in output in the less developed areas have been brought about mainly by increases in the cattle numbers; and increases in their yields are of a much smaller order relative to those in the developed areas. Although other reasons have contributed to these trends, it appears that technical progress has been more rapid in the developed countries.

Table 5 summarizes the picture for per caput food production.

TABLE 5. - INDEX NUMBERS OF PER CAPUT FOOD PRODUCTION, BY GROUPS OF REGIONS[1] (PREWAR = 100)

Area	Average 1948-52	Average 1957-61
Less developed regions (excl. China, Mainland)..	93	102
Less developed regions (incl. China, Mainland)..	88	100
Developed regions	110	127
World (excl. China, Mainland)	100	111
World (incl. China, Mainland)	97	109

[1] See note to Table 2.

As can be seen, per caput food production for the world as a whole regained the prewar level in the early 1950s and is now about 10% above this level. The improvement however has been largely in the more developed parts of the world. In these areas, per caput food production has risen by 27% above the prewar level. In contrast, in the less developed regions, it has barely exceeded the prewar level.

Table 6 brings out the striking disparity in the levels of per caput food production between the developed and the less developed regions.

TABLE 6. - LEVELS OF PER CAPUT FOOD PRODUCTION IN THE DEVELOPED AND THE LESS DEVELOPED REGIONS AS PERCENTAGES OF THE WORLD PER CAPUT AVERAGE LEVELS [1]

Area	Prewar	Average 1948-52	Average 1957-61
Less developed regions	64	58	59
Developed regions..........	174	194	203
World	100	100	100

[1] See note to Table 2.

Per caput food production in the developed regions was about 2.7 times as high as in the less developed regions during prewar years, compared with a ratio of 3.5 for the average 1957-61. This large disparity has further widened due to the increased share of the less developed regions in the world's population (from 67% to 72%) while their contribution to the world's food production has decreased from 43% to 42% (Table 7).

TABLE 7. - PERCENTAGE CONTRIBUTION OF THE DEVELOPED AND THE LESS DEVELOPED REGIONS TO WORLD FOOD PRODUCTION [1]

Area	Prewar	Average 1948-52	Average 1957-61
Less developed regions	43	41	42
Developed regions..........	57	59	58
World	100	100	100

[1] See note to Table 2.

Appendix 3 shows levels and trends in the index number of per caput food production, by regions. Food production per caput in Oceania is the highest in the world, four times the world average and eight times the level in the Far East. There is, however, no upward trend in the level of per caput production in this region. The level of per caput production in North America ranks only next to that in Oceania, being three times the world average. The level is now about 15% higher than before the war owing to the large expansion during the war and in the immediate postwar years, though more recently production per caput has levelled off because of restrictions on areas.

In both western and eastern Europe, per caput production is higher than the world average and has been growing at a faster rate. During postwar years increases were particularly high in eastern Europe, in fact higher than in any other region, mainly because of the rapid growth of production on largely expanded areas in the U.S.S.R.

In Latin America the per caput production level is close to the world average. After a considerable decrease during the war, the prewar level was approximately regained with the big expansion in output in the four years 1956 to 1959, but per caput food production has subsequently fallen back again to slightly below the prewar level.

In the other three underdeveloped regions, per caput production is lower than the world average by some 10% in the Near East, by more than 40% in Africa and more than 50% in the Far East. Among all the less developed regions, only in the Near East has per caput production been consistently maintained at more than the prewar level, whereas in each of the others this level seems to have been reached or exceeded sometime during the last decade. In Africa per caput food production increased above the prewar level and by the middle of the last decade was some 5% above it, but there has since been a decline. By contrast, food production per caput in the Far East in 1957-59 was still some 4% less than before the war, with little signs of improvement except possibly in the last two or three years. In Mainland China per caput production has been steadily increasing since the early 1950s, reaching its peak in 1958, the year of the "great leap forward," although the near-famine shortage of food reported recently resulting from the successive bad crops of 1959 and 1960 and the lack of stocks from previous years threw some doubts on the reliability of the high levels of per caput production reported for the late 1950s.

Food supplies

Trends in per caput food production do not fully reflect trends in per caput food supplies available for human consumption, because of external trade, changes in stocks and nonfood utilization.

Table 8 presents the index numbers of total food supplies per caput by regions. It also gives the index numbers of per caput food supplies of vegetable origin and of animal origin excluding oils and fats. Furthermore, the table summarizes the index numbers for two groups of countries, namely the low-calorie countries and the high-calorie countries. In keeping with the distinction made in the earlier world food surveys, the former comprise the Far East, Near East, Africa, and Latin America, excluding the River Plate countries, and the latter the remaining areas. The table shows that the world food supply per caput is some 6% above the prewar level compared with 9% in per caput food production. Part of the difference is explained by differences in time coverage and geographical coverage while part is due to accumulated stocks since prewar. The index numbers of per caput food supplies refer to the period 1957-59 and are based on available data for individual countries, while those of per caput production relate to regional estimates for 1957-61. The increase in the world per caput food supply level reflects mainly the improvements which have taken place in the high-calorie countries where per caput food supplies are some 20% higher than before the war. Those of the low-calorie countries, on the other hand, have barely exceeded the prewar level in spite of a large increase in net imports, resulting in further disparity in the supply levels of the two groups of countries. Thus, in prewar years the per caput supply level in the high-calorie countries was about 2.9 times as high as that in the

TABLE 8. - INDEX NUMBERS OF PER CAPUT FOOD SUPPLIES AVAILABLE FOR HUMAN CONSUMPTION, BY REGIONS [1] (Price weighted; prewar world average = 100)

Regions	Period	Crops, excl. vegetable oils and fats	Livestock and fish, excl. animal fats and oils	Total food (incl. vegetable and animal fats and oils)
Far East				
Incl. China, Mainland	Prewar	84	38	57
	Postwar	75	30	49
	Recent	83	38	56
Excl. China, Mainland	Prewar	80	35	53
	Postwar	69	27	44
	Recent	80	35	53
Near East	Prewar	112	72	90
	Postwar	108	71	88
	Recent	125	72	102
Africa	Prewar	...	...	...
	Postwar	...	...	...
	Recent	87	44	65
Latin America	Prewar	77	123	95
	Postwar	86	97	88
	Recent	93	110	99
Europe (incl. U.S.S.R.)	Prewar	120	154	141
	Postwar	120	147	137
	Recent	120	199	168
North America	Prewar	178	394	313
	Postwar	172	460	339
	Recent	157	495	351
Oceania	Prewar	181	355	284
	Postwar	194	346	283
	Recent	179	354	282
Low-calorie countries [2]	Prewar	84	47	61
	Postwar	78	39	55
	Recent	87	46	62
High-calorie countries	Prewar	131	205	177
	Postwar	132	220	184
	Recent	128	269	211
World [2]	Prewar	100	100	100
	Postwar	94	95	95
	Recent	99	112	106

[1] These index numbers have been calculated by applying regional weights, based on 1952-56 farm price relationships, to estimated regional per caput supplies of major food groups.
[2] Includes estimates for Africa for prewar and postwar.

low-calorie countries while the disparity is now 3.4 to 1.[1] Levels are highest in North America and Oceania, more than 3 times the world average and 6 to 7 times the low level in the Far East.

The table also shows that the increase in the index number of per caput food supply for the world since prewar is due to an increase of some 12% in the food supplies of animal origin while those of vegetable origin are still slightly below the prewar level. However, the increase in the index number of foods of animal origin took place in the high-calorie countries; the figure for the low-calorie countries remains below the prewar level. In contrast the index number of food supplies of vegetable origin remained relatively stable with a small decrease in the high-calorie countries and a similarly small increase in the low-calorie countries. In other words, the quality of the diet improved only in the high-calorie countries while the low-calorie countries were barely able to maintain the quantity of the diet.

The table brings out strikingly the large disparity between the two groups of countries in the levels of supplies of foods of animal origin against a relatively small difference in those of vegetable origin. The former are 6 times higher in the high-calorie countries while the latter are only 1½ times higher compared with the low-calorie countries.

The disparities between regions in per caput food supplies are best brought out in Table 9 which shows region by region the share in the world's population against the corresponding share in food supplies.[2] The shares are shown separately for food supplies of vegetable origin and of animal origin.

Over half of the world's population in the Far East is seen to live on only about a quarter of the world's total food supplies made up of only 19% of the world's animal food supplies and 44% of the world's crop food supplies. This is in striking contrast with the position in Europe, Oceania, and North America which with a share in population of 29% account for 57% of the total world's food sup-

---

[1] The figure is not directly comparable with the figure of 3.5 to 1 for the disparity in per caput food production found in the previous section, due to the different geographic and time coverages. The same holds for a comparison of Tables 9 and 7.

[2] See footnote 1.

TABLE 9. - DISTRIBUTION OF WORLD POPULATION AND FOOD SUPPLIES, BY REGIONS (1957-59)

Regions	Percentage of population	Percentage of food supplies		
		Total	Animal	Crops
Far East (incl. China, Mainland)	52.9	27.8	18.5	44.2
Near East	4.4	4.2	2.8	5.5
Africa	7.1	4.3	2.8	6.3
Latin America	6.9	6.4	6.7	6.5
Europe (incl. U.S.S.R.)	21.6	34.2	38.4	26.2
North America	6.6	21.8	29.2	10.4
Oceania	0.5	1.3	1.6	0.9
World	100.0	100.0	100.0	100.0

plies, 69% of the total animal food supplies, and 38% of the total crop food supplies.

In Africa the ratio of the shares of population and food supplies is 1.7:1 for total food supplies and 2.5:1 for animal food supplies. In the Near East it is 1:1 for total food supplies and 1.6:1 for animal food supplies while in Latin America it is 1:1 for both.

Underlying economic and social factors

The disparities in per caput food supplies between the developed and less developed regions are, to an appreciable extent, associated with the disparities in per caput food production which are in turn associated with the disparities in various measures of agricultural productivity, in farm incomes, and in over-all population densities. Table 10 illustrates the disparities in the levels of production per unit of land and of cattle. Yields in the developed regions for some of the crops are seen to be nearly twice as high as in the less developed areas and the disparities in the output of meat and milk per unit of cattle is even 5 to 1. This is of course partly due to the preponderant practice in the less developed regions of using cattle as draught animals (and sometimes by keeping cattle for other reasons than production), but even when this factor is allowed for, the disparities in

TABLE 10. - TRENDS IN YIELDS OF MAJOR CROPS AND CATTLE PRODUCTS,[1] BY GROUPS OF REGIONS

Commodity	Less developed regions			Developed regions			World		
	Prewar	Average 1948-52	Average 1957-60	Prewar	Average 1948-52	Average 1957-60	Prewar	Average 1948-52	Average 1957-60
	100 kg/ha								
Crops									
Wheat	9.2	7.9	9.4	10.3	11.2	13.3	9.9	10.0	11.9
Rice, paddy	17.5	15.9	19.1	32.1	28.1	38.4	17.6	16.1	19.2
Other cereals	8.6	7.3	9.0	12.3	13.9	18.0	10.7	10.7	13.3
Starches	67.8	68.0	74.5	108.5	114.9	122.2	96.9	95.3	97.1
Pulses	6.5	6.1	6.1	6.3	5.7	6.5	6.5	6.0	6.2
		Average 1948-52	Average 1958-60		Average 1948-52	Average 1958-60		Average 1948-52	Average 1958-60
	100 kg/head of cattle								
Cattle Products		2.5	2.6		12.3	13.9		5.7	6.5

[1] Meat and milk in terms of milk equivalent, taking 1 unit of meat as equal to 10 units of milk.

average yields are still high. The table shows also that the disparities in yield have tended to widen.

The reasons for the low productivity of labor in the less developed countries are usually clear enough. Food production is often dispersed among a multitude of farms most of which are very small where the output may barely provide a minimum subsistence for the family, any excess being only sufficient for extremely modest cash requirements. But even on small holdings farmers can try to compensate for their disadvantages by striving for a high output per hectare. For instance in Japan where over 60% of the holdings are under 1 hectare, yields are as high as anywhere else in the world. The laborious technique of cultivating rice and other crops in Japan depends on the industry of the Japanese peasant. But there are other important factors. The rainfall is adequate and the irrigation system almost perfectly controlled. Farmers have generally avoided wasting land on draught animals and, more recently, have obtained mechanical equipment suited to small holdings. Large urban markets, supplies of cheap industrial goods, and security of tenure have all provided economic incentives to increased production. The thrift of the Japanese peasant, the high general level of education, and a well-developed co-operative system may all have provided a favorable background against which the Japanese achievement became possible.

It is known, however, that in most countries where farms are very small, the productivity of land is low. This is particularly true where farm practices are backward and credit facilities to acquire essential farm requisites are mainly lacking, as in most of the less developed countries. In many cases the problem is further aggravated by inadequate supplies and distribution of irrigation water. In some areas rainfall is of such great intensity and soil conservation measures are so inadequate that soil is often washed away. Furthermore, in many areas, and particularly in the Far East, little has been done over the years to restore the fertility of the cultivated soil. Illiteracy is widespread and most farmers know little about improved methods and techniques of production. Organizations for technical services to farmers are inadequate or nonexistent. Even where such organizations are established farmers have often been unwilling to adopt the improved practices partly because of attachment to traditional ways. They may be reluctant to adopt new methods where this involves

TABLE 11. - PER CAPUT NET PRODUCT: TOTAL, AGRICULTURAL AND NONAGRICULTURAL; PERCENT CONTRIBUTION OF AGRICULTURE TO DOMESTIC PRODUCT AND PERCENTAGE OF AGRICULTURAL POPULATION IN VARIOUS COUNTRIES (1952-54)

Countries	Per caput net product at factor cost			Percentage of per caput agricultural to nonagricultural income	Percent contribution of agriculture to domestic production	Percentage of agricultural population
	Total	Agricultural	Nonagricultural			
	.... *U.S. dollars*					
Canada	1 310	983	1 372	72	12	16
U.S.A.	1 870	668	2 065	32	5	14
Australia	950	1 306	882	148	22	(16)
New Zealand	1 000	1 091	974	112	24	(22)
Belgium	800	492	846	58	8	13
Denmark	750	656	780	84	21	(24)
France	740	455	840	54	16	(26)
Germany, Fed. Rep. of	510	364	534	68	10	(14)
Sweden	1 160	706	1 296	54	14	23
United Kingdom	780	557	797	70	5	(7)
Argentina	460	350	496	71	19	25
Brazil	230	110	435	25	30	(63)
Ecuador	150	97	237	41	40	(62)
Paraguay	140	128	156	82	51	56
Congo (Leopoldville)	70	27	165	16	27	(69)
South Africa	300	145	376	39	16	33
India	60	41	104	39	48	70
Japan	180	106	243	44	27	46
Pakistan	70	54	120	45	59	76
Philippines	150	91	295	31	43	(71)
Thailand	80	58	122	48	48	66
Turkey	190	109	380	29	40	70
U.A.R.	120	67	211	32	35	63

SOURCE: FAO, *The state of food and agriculture 1959*, Rome, Italy.
Note: Figures in parentheses refer to percentage of total male workers engaged in agriculture.

higher cost, more investment, and more contact with the marketing system; especially so in a situation in which land tenure is insecure, and the basic services of marketing, transport, and storage are bad or unreliable. Many of the changes the farmer may be asked to make will increase the risk that his already inadequate income will dwindle or disappear completely.

Where the majority of farms are small and agriculture is the predominant economic activity, inevitably a large part of the population is dependent on agriculture. As Table 11 shows, in the less developed countries well over one half of the population is dependent on agriculture. In these countries a farm family barely produces the food that it needs for itself and there is at best only one nonfarm family per farm family to which to sell its additional produce. Furthermore, the production is confined to few commodities. Consequently the diets, consisting mainly of staple foods, lack variety. In contrast, in many of the developed countries the population dependent on agriculture is small, often less than 20%. Besides, the level of productivity in the developed countries is so high that one farm family produces much more than the quantity of food it needs. Furthermore, there are 10 to 20 nonfarm families on the average to buy its produce. Sales of home-grown foods enable the farm family to purchase types of foods not produced on the farm. Thus even in rural areas, much greater variety in food supplies is available.

The low level of productivity of labor in agriculture is a main cause of the low general level of incomes in less developed countries as illustrated by Table 12.

Although 70% of the world's people live in less developed areas their contribution to world income is little over 20%. In fact,

TABLE 12. - DISTRIBUTION OF THE WORLD'S POPULATION AND OF NATIONAL INCOMES, BY REGIONS (1957-59)

Regions	Percentage of population	Percentage of income
Far East (incl. China, Mainland)	52.9	12.3
Near East	4.4	1.9
Africa	7.1	2.5
Latin America	6.9	4.8
Europe (incl. U.S.S.R.)	21.6	39.5
North America	6.6	37.8
Oceania	0.5	1.7
World	100.0	100.0

per caput annual income early in the last decade was less than $100 in most countries in the Far East and Africa; $100-200 in most countries of the Near East, and $100-250 in most countries of Latin America. Against this, per caput income was between $750-1000 in western European countries, around $1000 in Oceania, and $1500-2000 in North America (see also Table 11).

These disparities have not been reduced in the course of the last decade but rather the gap has widened. Income in the less developed countries of the world grew at 3% per annum in the last decade. Because of the growth of population the increase in income per person was only $1 per annum. In contrast the average income in the developed countries during the same period rose by about $20 per head per year. This widening gap is also due to increasing disparities between the export prices of manufactured products and related services and those of primary products. The less developed countries in general are earning less and less per unit exported primary domestic product and are paying more and more per unit imported industrial product. Furthermore, in their efforts to accelerate economic development many less developed countries require more foreign currency earnings. This in turn leads to more concentration on export cash crops most of which are nonfood products; contributing to lower per caput food production and consequently, owing also to the need to limit imports, to lower per caput food supplies.

Low as the per caput income is in the less developed countries, even lower is the per caput farm income (Table 11). In the more developed countries also, disparities between income from farming and income from other activities are still to be found, though to a much lesser extent. As a consequence, rural-urban disparities in per caput food supplies are wider in the less developed than in the developed regions.

In the more developed countries food consumption has reached levels where further rises in disposable income lead to only relatively small increases in expenditure on food. The effect on farm sales of food products is still less, since part of any increased expenditures on food reflects more elaborate methods of processing and distribution. Thus the growth of demand for farm products is hardly greater than the growth of population which, although fairly rapid in North America and Oceania, is very slow as compared to the less developed areas. Agricultural production, however, tended to increase rapidly in the developed countries, for technical progress has been substantial and its adoption in farm practice was encouraged by agricultural support policies. These circumstances have led recently to surplus food production accumulating in some of these countries.

In contrast, in the less developed countries it is more often the demand for food that runs ahead of the supply. Population growth has been fast and any increase in the low levels of income was largely spent on food and other agricultural products. Because production has fallen behind demand, domestic food prices have tended to rise and many less developed countries have had to import more food or to restrict exports of foodstuffs.

With the low level of productivity in less developed areas and with the low incomes, savings are low and offer small possibilities of investment to improve productivity or for new exploitations. Low income, low food consumption, and low productivity thus go together in a "circle of poverty" which is still common to nearly all the less developed countries.

3. LEVELS, PATTERNS, AND TRENDS OF FOOD CONSUMPTION

Major food groups

Appendix 4 gives the data on per caput food supplies of major food groups by regions and subregions for prewar, postwar, and recent periods. The salient features are summarized in the following paragraphs.

Cereals

Among the regions, consumption of cereals is highest in the Near East, exceeding 160 kg per caput per year, followed by the Far East with 146 kg, and is lowest in North America and Oceania with about 70 kg and 90 kg, respectively. Consumption levels in Africa and Europe are close to the world average of 134 kg. The level in eastern Europe (including the U.S.S.R.) is higher than in the rest of Europe; in fact it is among the highest in the world.

The pattern of consumption of cereals is different from region to region and within regions. Thus, in the Far East rice is usually the preferred cereal, to the extent that the word is sometimes used as synonymous with food. Exceptions are to be found in north China and parts of India and Japan where wheat, maize, and millets are more important, while in parts of Indonesia maize is the common cereal. The Near Eastern countries depend heavily on wheat and to a lesser extent on millets, maize, and barley. North African countries, like the other Mediterranean countries, rely heavily on wheat and barley as their staple foods. In the savannah areas of Africa, millets are most important, followed by maize. Substantial quantities of rice are consumed in certain coastal zones. In east Africa the pattern is diverse, though maize is the principal crop. In Latin

America maize is the staple food except in the plains to the south where wheat is more important. The tropical coasts and plains are the rice-growing areas.

In the postwar period, per caput consumption of cereals compared with the prewar level had declined everywhere except in Latin America where it was some 10% above the prewar level. During the period from prewar to postwar years the Far East and the Near East regions, which had previously been net exporters of cereals, became net importers. Following this period of shortages, per caput consumption in the Far East and Latin America has been rising during the last decade, and in the Near East it was maintained at the high postwar level. This was due amongst other factors to the increasing imports on concessional terms from the larger exporting countries where considerable stocks of wheat and coarse grains have been accumulating. In the developed regions on the other hand the declining trend has continued over the last decade.

Starchy roots

Per caput consumption is highest in west and central Africa, nearly four times the world average of 84 kg per year, mainly due to the high consumption of cassava and yams. However, the data for Africa should be interpreted with more care since estimates of production of cassava may in many cases refer to potential production rather than to the quantities actually harvested for consumption. Next to west and central Africa is the consumption level of eastern Europe (including the U.S.S.R.) with 179 kg per caput per year, mainly potatoes, followed by the major islands of southeastern Asia with 136 kg, mostly cassava and yams. Consumption is lowest in the Near East, being only about 20% of the world average.

During the postwar years per caput consumption for the world as a whole increased by some 10% because of increases in Europe, Oceania, Latin America, and the Far East; in the other regions consumption decreased. During the last decade, per caput consumption for the world as a whole increased by a further 10%. There was a sharp increase in consumption in the Far East, where the need for calories is greatest, and a modest increase in the Near East, as also in Oceania and possibly the U.S.S.R., whereas in other regions consumption levels tended to decline. The decline was sharpest in Europe (excluding U.S.S.R.) and North America, where the level fell below the prewar average.

SUGAR

Latin America is the only less developed region in which the consumption of sugar is above the world average of 18 kg per caput per year; in fact, the consumption is nearly twice this figure, and some 15% higher than in Europe. Per caput consumption is highest in Oceania and North America, 3-4 times the world average, and lowest in the Far East, where the level is about half the world average.

From prewar to postwar periods, only Latin America recorded a substantial increase of about 20%, while for the world as a whole the per caput consumption remained unchanged. During the 1950s, world per caput consumption of sugar rose by 2-3% per year. The rise in consumption during the last decade was partly due to the fact that sugar prices have developed more favorably than those of any other food crop. Trends in per caput consumption differed widely between the various regions. In North America and Oceania, consumption remained relatively stable at a high level. In the other regions, per caput consumption went up steadily, the greatest increases being recorded in eastern Europe and the northern countries of Africa.

In all the less developed regions consumption is now higher than in prewar.

PULSES AND NUTS

Consumption of pulses and nuts in the less developed regions amounts to 18 kg per caput per year. This is about 3 times as high as the level in Europe, 2½ times as high as that in North America and 4½ times that in Oceania.

Pulses are important in the diets in the Far Eastern countries, especially in southern Asia in the form of gram and in eastern Asia in the form of soybean and soybean products, while in the semitropical areas of southeastern Asia coconuts are more important. Pulses are a common food in most Latin American countries. Because of the low intake of animal protein, the consumption of pulses is a desirable feature of the diet in these areas due to their high protein content which supplements that from cereals.

In prewar years the consumption level was highest in the Far East, but it declined in postwar years and continued to decrease during the last decade, though to a lesser extent. Also in the Near East and Latin America consumption levels declined between prewar and postwar, but they increased rapidly during the last decade and in fact have now reached the level of the Far East. In contrast, in Africa the consumption level increased between prewar and postwar but has remained stable since then. In the developed countries, consumption levels have varied little. For the world as a whole, per caput supplies have remained almost unchanged during the last decade and are slightly below the prewar level.

Vegetables and fruit

Of all foods, the data for vegetables and fruit are most difficult to estimate. The available information indicates that the consumption level is lowest in the Far East, about $^3/_5$ of the world average of some 90 kg per caput per year. In this region fruits are usually considered luxuries, except in tropical and semitropical areas where they are often eaten directly from the bush. Fresh green and yellow vegetables which are so valuable as a source of important vitamins and minerals are consumed in very small quantities. Preserved vegetables are more common and can provide a good source of vitamins and minerals. In Africa, the consumption is highest in the northern countries where it approximates the world average. In all other regions the consumption is well above the world average, the maximum being in North America. Among the less developed regions the Near East has the highest consumption, about the same as in Oceania and higher than in Europe.

From prewar to postwar periods, there was an upward trend in per caput consumption in all the regions except the Far East where it declined some 15%; and for the world as a whole it remained practically unchanged. During the last decade there was an upward trend in Europe and in the less developed regions, though in the Far East the level still remains nearly 10% below the prewar average. On the other hand, Oceania and North America registered a decline during the same period, with the result that consumption is back to the prewar level in Oceania and a little below prewar in North America. For the world as a whole there has been an increase of about 10% above the prewar level.

Milk and milk products (excluding butter)[1]

The average consumption throughout the world is equivalent to 82 kg of liquid milk per caput per year; but the differences in consumption are very wide. This is especially important in relation to the adequacy of diets of expectant and lactating mothers and of children. Consumption in the Far East is only a quarter of the world average (in fact it is substantially lower than this because milk going into ghee (butterfat) is included in this group in India and Pakistan); in Africa the figure is considerably higher, but consumption is low in west and central Africa. Consumption in Europe is over twice the world average and in North America nearly 4 times the world average (15 times the consumption in the Far East).

Generally speaking, consumption levels in developed and less developed regions have been moving in opposite directions and the disparities have been getting wider.

Meat, fish, and eggs

The differences in consumption in this group between developed and less developed regions are nearly as striking as those for milk

[1] There is no correct way of converting quantities of milk products which have been made from 'incomplete' (i.e., partly skimmed milk) into whole milk equivalent. For practical purposes pragmatic conversion factors have been used. This seems to be justified by the fact that conversion of milk and milk products to fat and protein equivalents in most countries shows about the same quantities of both components as whole milk.

and milk products. The world average is about 33 kg per caput per year; the contribution of fish and eggs to this total being about 15% and 12%, respectively. Certainly this proportion is much more variable for fish, consumption being determined largely by ready access to supplies. Consumption of meat, fish, and eggs is less than half the world average in the Far East, 3½ times the world average in North America. In Latin America the consumption is well above the world average, but this is almost entirely due to the remarkably high level in the River Plate countries.

Except in North America, meat consumption fell from the prewar to the postwar period; generally the drop was compensated by a rise in fish and eggs. Recently, world meat consumption has regained its prewar level (but consumption is still short of this level in Latin America). Most low-calorie countries have shared in a modest way in the increase in meat since the postwar period. Since the war, fish consumption has continued to rise in the less developed regions, but eggs have only risen in Europe and Latin America.

Fats and oils

Per caput consumption in the more developed regions is 16-21 kg per year. This is 2 to 7 times the levels in the less developed regions.

Trends between prewar and postwar were generally in the opposite direction in the two groups of regions with supplies declining in the more developed regions. During the last decade consumption levels have increased in all the regions except in the Far East where the recent level has fallen back below the prewar average. For the world, consumption has remained unchanged since prewar.

Summing up, it is clear that whereas the per caput consumption of foods rich in carbohydrates has been nearly as high in the less developed regions as in the developed regions, consumption of animal products and other protective foods including fats and oils in the less developed areas is far below that in the developed regions.

During the last decade, in the less developed areas consumption of carbohydrate foods and also to some extent of animal products (and other protective foods), has increased although the latter is still slightly below the low prewar level. In the developed regions, on the other hand, the consumption of carbohydrate foods has declined but that of animal products has increased substantially.

Calories and proteins

The disparities in the per caput food supplies are best brought out in terms of the nutritive value of the diets. This has two broad aspects, quantitative and qualitative. The former is measured by the calorie content of the diet, but to assess the latter no single and simple measure is available, since the balance between many nutrients needs to be taken into account. However, two indicators of the nutritional quality of the diet have been widely accepted, namely its protein content, and in particular the animal protein content, and the percentage of total calories derived from cereals, starchy roots, and sugar.

Table 13 shows the average levels of calorie and protein supplies for the prewar, the postwar, and the recent periods. The table shows that the less developed regions consume between 2,000 to 2,500 Calories per caput per day. The food consumed by the more developed regions, on the other hand, provides more than 3,000 Calories per caput per day. The contrast between the two groups of regions is seen to be even more striking in respect of the nutritional quality of the diets. Thus, the less developed regions consume 9 g of animal protein per caput per day, which is only one fifth the consumption per head in the more developed regions. Again, the percentage of calories derived from cereals, starchy roots, and sugar is seen to be close to 80 in the less developed regions against less than 60 in the developed regions. The disparity in the supply of fats is also large, levels being a little over 30 g per caput per day in the less developed regions and over 100 g in the developed regions.

It appears that the current level of calorie supplies has reached, and even exceeded, the prewar level in all developing regions except the Far East where it is still somewhat lower than the prewar level. Also in Europe, the current level is higher than the prewar level,

TABLE 13. - AVERAGE DAILY PER CAPUT SUPPLIES OF CALORIES, TOTAL PROTEINS, AND ANIMAL PROTEINS AT THE RETAIL LEVEL, BY REGIONS

Regions	Period	Calories		Total protein	Animal protein
		Number	% derived from cereals, starchy roots, and sugar		
				..........Grams..........	
Far East (incl. China, Mainland)	Prewar	2 090	78	61	7
	Postwar	1 890	79	54	6
	Recent	2 060	81	56	8
Near East ...	Prewar	2 295	78	72	12
	Postwar	2 220	78	69	12
	Recent	2 470	71	76	14
Africa	Prewar	...	...	...	...
	Postwar	...	...	...	...
	Recent	2 360	74	61	11
Latin America	Prewar	2 160	63	64	28
	Postwar	2 315	66	62	22
	Recent	2 510	63	67	24
Europe (incl. U.S.S.R.)	Prewar	2 870	67	85	28
	Postwar	2 760	68	82	29
	Recent	3 040	63	88	36
North America	Prewar	3 260	48	86	51
	Postwar	3 170	43	91	61
	Recent	3 110	40	93	66
Oceania	Prewar	3 290	50	103	67
	Postwar	3 250	50	98	66
	Recent	3 250	48	94	62
Low-calorie countries [1] ..	Prewar	2 110	77	62	10
	Postwar	1 960	78	56	8
	Recent	2 150	78	58	9
High-calorie countries ...	Prewar	2 950	62	85	34
	Postwar	2 860	62	85	37
	Recent	3 050	57	90	44
World [1]	Prewar	2 380	71	69	18
	Postwar	2 240	71	64	18
	Recent	2 420	70	68	20

[1] Includes estimates for Africa for prewar and postwar.

but there has been a declining trend in North America and Oceania. For the world as a whole calorie supplies would appear to have just reached the prewar level. Total protein supplies for the world as a whole per caput per day have gone up since the early postwar period but have still not reached the prewar level. The gap remains particularly large in the Far East. Animal protein supplies for the world as a whole have exceeded the early postwar as well as the prewar level. This however is due to considerable increase in consumption in Europe and North America. On the other hand, the animal protein supplies for the underdeveloped regions have barely reached or are still below the prewar level, thus further widening the already large gap between the developed and the less developed regions. Similar trends are revealed by the percentage of calories derived from cereals, starchy roots, and sugar.

4. HUNGER AND MALNUTRITION

Undernutrition means inadequacy in the quantity of the diet, that is, in calorie intake which, continued over a long period, results in either loss of normal body weight or reduction in physical activity, or both. This definition is strictly appropriate to adults, not to children. For children, the consequences of low calorie intake are unsatisfactory growth and physical development and a reduction of the high degree of activity characteristic of healthy children.

Malnutrition means inadequacy of the nutritional quality of the diet which if made good enables a person to lead a healthy active life. More precisely, it denotes inadequacy of a particular or several essential nutrients. Serious shortages of nutrients may result in clinical signs of specific deficiency diseases; minor degrees of deficiency can contribute to poor general health.

Undernutrition and malnutrition are naturally not mutually exclusive: people who are undernourished are likely to be malnourished and this, particularly if serious or prolonged, will lower resistance to disease.

Undernutrition

Evidence of undernutrition

Table 14 compares region by region the existing levels of calorie supplies with the corresponding requirements, shown to the nearest 50. It shows that the calorie supplies for the Near East, Africa, and Latin America are about equal to their requirements, those for the Far East fall short of the requirements by about 10%, while those for Europe, North America, and Oceania are not only sufficient to meet their average needs but in fact exceed them by about

20%. These gaps between the supply and requirement levels need, however, to be interpreted with care, since within regions the calorie intake as well as the requirement varies considerably. For example, the apparent self-sufficiency of calorie supplies in the Near East is due to the somewhat higher level of supplies in Turkey, the United Arab Republic, Syria, Lebanon, and Israel which account for about one half the population in the region. The other countries in the region, which include Iran, Iraq, Saudi Arabia, and Jordan, have Calorie supplies only of between 2,100 and 2,200 per caput per day. The same observation applies to the Far East where, for example, Japan and China (Taiwan) are relatively much better fed than other countries in the region. Within-country variation between different socio-economic classes of the population is even larger in the less developed countries. The table brings out the large variation in calorie supply between regions against the relatively small variation in calorie requirements. Thus, while admittedly the calorie needs of the people in the less developed regions are smaller than those of the people in Europe and North America, as wide a difference as 850 Calories between the consumption levels of the two can hardly be justified on grounds of differences in age distribution, stature, and climate.

TABLE 14. - LEVELS OF CALORIE SUPPLIES AND CALORIE REQUIREMENTS BY REGIONS, 1957-59 (PER CAPUT PER DAY AT RETAIL LEVEL)

Regions	Calorie supplies [1]	Calorie requirements [1]	Calorie supplies as percentage of calorie requirements [2]
Far East (incl. China, Main.)	2 050	2 300	90
Near East	2 450	2 400	103
Africa	2 350	2 400	98
Latin America	2 500	2 400	104
Europe (incl. U.S.S.R.)	3 050	2 600	117
North America	3 100	2 600	120
Oceania	3 250	2 600	125
Low-calorie countries	2 150	2 300	93
High-calorie countries	3 000	2 600	116
World	2 400	2 400	100

[1] Rounded to the nearest 50.
[2] Calculated on unrounded figures.

In particular, the foregoing observation holds for the Far East, where a gap of 10% is to be found between the average calorie supply and the requirement. The significance of this gap cannot be overstressed. Considered in terms of total supplies the gap is seen to be large enough to feed the entire population of the Near East. The gap may either persist all through the year or be more severely felt in the period preceding the harvests. Again, it may be shared by the majority of the population or fall heavily on the poorer sections. In actual fact, the privileged and well-to-do everywhere will eat all they need and perhaps more. But what the poor can afford will generally not meet their full needs. Data from household surveys conducted in India showed that in practically every part of the country there are a large number of families who for want of income live on quantitatively inadequate diets and who do not get sufficient quantities even of staple food grains (Table 15). The data for India are illustrative of the conditions in other less developed countries.

By contrast, in the developed countries there is on the average no shortage of calories even in the poorest classes. Relevant data from a nation-wide survey conducted in the United States illustrate this statement. Table 15 shows that calorie supply is not influenced by rising income on anything like the scale noticed in India. Further, the level of calorie supply even for the lowest income groups in the United States is higher than the average requirement.

TABLE 15. - PER CAPUT INCOME AND FOOD CONSUMPTION

United States, 1955		India (Maharashtra State), 1958	
Yearly household disposable income in U.S.$	Food consumption Calories/caput/day [1]	Yearly per caput expenditure in U.S.$	Food consumption Calories/caput/day [1]
under 3 000	3 200	under 33	1 500
3 000-5 999	3 150	33-59	2 300
6 000 and over	3 250	60 and over	2 900

[1] Rounded to the nearest 50.

INCIDENCE OF UNDERNUTRITION

What has already been stated suggests that a large number of people in the low-calorie countries go undernourished at least part of their lives. This information, however, is not sufficient to estimate the proportion of people who are undernourished. To determine this proportion requires a comprehensive knowledge of the distribution of calorie intake in relation to the requirements of individuals. Such data are, however, scarce. What are available are the calorie intake distributions of households on "reference-man" basis[1] such as are given in Table 16 for prepartition India and for Burma.

TABLE 16. - DISTRIBUTION OF HOUSEHOLDS SURVEYED IN PREPARTITION INDIA (1935-48)[1] AND BURMA (1939-41) BY CALORIE SUPPLIES PER DAY PER REFERENCE MAN

Calories per day per reference man	Percentage frequency of households	
	India	Burma
under 1 300	4.9	.3
1 300 - 1 700	10.7	5.2
1 700 - 2 100	23.2	20.3
2 100 - 2 500	21.8	29.4
2 500 - 2 900	20.7	23.9
2 900 - 3 300	12.0	10.4
3 300 - 3 700	4.2	5.7
3 700 - 4 100	1.3	3.3
4 100 - 4 500	.7	1.0
4 500 and over	.5	.5
	100.0	100.0

[1] Refers to average consumption in small groups of households (see text, p. 41).

A method has been developed for estimating the proportion of undernourished households based on such distributions, using the available knowledge of the variation in energy expenditure among young, healthy, active adults, and is illustrated below with reference to the data for Burma.[2]

---

[1] The term "reference man" is a basic concept of the FAO calorie-requirement scale. The calorie requirement of the reference man for any country is that of a healthy, adult male of 20-30 years of age of moderate activity with an assumed desired weight, adjusted for the average environmental temperature.

[2] For details and sources of data see Sukhatme, P. V., "The world's hunger and future needs in food supplies," *The Journal of the Royal Statistical Society*, Series A (general), Vol. 124, 463-525, 1961.

The distribution for Burma refers to dietary surveys conducted during 1939-41 in representative areas of the different districts of the country. In each selected area 20-30 households representing the population preponderant in the area were studied. The data were collected by local trained volunteers who were known to the households visited. The records of the quantity and kind of foodstuffs consumed were obtained by two daily visits at the time of preparation for cooking the principal meal. The foodstuffs were weighed before preparation. Sex and age of all members of the household consuming food were noted, and in calculating the reference-man equivalent of the members of the households, notice was also taken of any absence from meals or the presence of any guests. The table shows that in two thirds of the surveyed households in Burma the calorie intake per reference man falls short of the corresponding requirement, which according to the FAO scale is approximately 2,700. This does not mean, however, that two out of every three households are underfed, any more than one out of every four households found to fall short of the corresponding requirement in the United States (Table 18) can be considered undernourished. Clearly, some people will need less than the stipulated requirement, while others may need more. This is so because the adjustment of individual intake to reference-man equivalent is based on age and sex only and disregards variations in physical activity and other residual factors.

Available evidence given in FAO's report on calorie requirements indicates that the standard deviation of energy expenditure can be placed at approximately 400 Calories per reference man.[3] The value of the standard deviation for households with a size equivalent to four reference men as is the case in Burma can be placed at approximately 200. If Burma's population were adequately nourished one would expect only a very small fraction of the households to have Calorie intake on a reference-man equivalent basis below 2,700 less three times the value of the standard deviation, that is, below 2,100. In

[3] This figure is in broad conformity with the findings of Harries, J. M., Hobson, E. A. and Hollingsworth, D. F., " Individual variations in energy expenditure and intake," *Proceedings of the Nutrition Society*, Vol. 21, 157-169, 1962.

actual fact, the proportion of households with intake below 2,100 Calories is approximately 25%; assuming that the sample is representative, a conservative estimate of the proportion of undernourished in Burma during the period of the survey would be roughly a quarter.[4]

The data for prepartition India tell much the same story. These were collected over a 14-year period from 1935 to 1948, were fairly evenly spread out throughout the different seasons and covered 843 groups totalling 12,500 households, mainly drawn from the low-income stratum, estimated to cover about two fifths of India's population. The standard deviation appropriate for estimating the incidence of undernutrition from this distribution is approximately 100 Calories. It would therefore appear that most households would need to take more than 2,400 Calories per reference man in order to meet their needs. In actual fact, as Table 16 indicates, 55% of the groups of households surveyed during 1935-48 took less than 2,400 Calories. This is equivalent to approximately one fifth of all households in India. There would undoubtedly be some households with inadequate calorie supplies in the middle income group of the country, but no estimate of the proportion of such households can be made. Generalizing from the above, it would appear that the proportion of undernourished in India during 1935-48 was about one fourth.

A similar analysis of a survey conducted in Ceylon (1948-49) indicates that the proportion of undernourished was a little over one third.

It may be argued that all these data refer to earlier years and that the consumption levels have since probably improved. This however does not appear to be the case. Table 17 shows the average daily intake of foodstuffs per reference man as revealed by dietary surveys conducted by the Indian Council of Medical Research dur-

---

[4] It is recognized that food is not equitably distributed in proportion to the needs of individual members, especially in poor households. Earners, for example, might take an adequate share of the food, leaving children to feed on much less than what they need. On the other hand, the same may also hold, although to a somewhat smaller extent, for households with more than 2,100 Calories per reference man. In generalizing the conclusion regarding the proportion of undernourished households to undernourished populations, it is assumed that these two groups broadly counterbalance each other.

ing 1935-48 and those carried out during 1955-58. Although the households surveyed during 1955-58 were not the same as those surveyed in the earlier period, the two groups are stated to be broadly comparable and, further, each included a large number of households selected from different parts of the country. The table shows that there has been no appreciable quantitative change in the diet of the people in India during the last 15 years. The estimate of between one third (Ceylon) to one fourth (India and Burma) regarding the proportion of undernourished households would thus seem to hold good even today.

TABLE 17. - AVERAGE INTAKE OF FOODSTUFFS PER REFERENCE MAN AS REVEALED BY FOOD CONSUMPTION SURVEYS IN INDIA

Commodities	1935-48 [1]	1955-58 [2]
	Ounces per day	
Cereals	16.6	16.6
Pulses	2.3	2.4
Leafy vegetables	0.9	0.7
Other vegetables	4.1	3.2
Ghee and oils	0.9	0.5
Milk	3.3	2.8
Meat, fish and eggs	0.9	0.5
Fruit (and nuts)	0.6	0.2
Sugar and jaggery	0.7	0.7

[1] Data refer to prepartition India.
[2] Data refer to India only.

The limited statistics from other underdeveloped countries in the Far East lead to much the same conclusion as for India, Pakistan, Ceylon and Burma which comprise one third of the population of the Far East. Thus the conclusion is probably valid, with small differences, for the whole region, except Japan and China, Taiwan, where the calorie supply is relatively higher.

However, the factor which is likely to make the largest difference to the estimate of the proportion of the world's people who are underfed is the estimate for Mainland China. The food balance sheet for Mainland China for 1957-59 prepared by FAO shows that the Calorie consumption level for this country is around 2,100 as against the consumption level of around 2,000 for India and Pakistan. On the other hand, the requirement for Mainland China is also likely to be somewhat higher as compared to that of India and Pakistan, in view of the more intensive use of labor in that country and lower environmental temperatures. Even granting that conditions in Mainland China have improved substantially in the course of the last 10 years, the recent famine in the country (suggesting a lack of reserves from the exceptionally good crops reported for the previous years) indicates that consumption in Mainland China has not yet reached a level where the existence of undernutrition on an appreciable scale even in normal times can be ruled out. Granted that the proportion of undernourished in Mainland China is somewhat smaller than in the subregion comprising India, Pakistan, Burma, and Ceylon, it is nevertheless likely to be appreciable and, possibly, amounts to 20%. It would therefore seem that about one fifth of the population in the Far East and possibly more is undernourished. We have already seen that though the over-all average calorie supply in other underdeveloped regions is about equal to the respective over-all requirements, there are a number of countries in these regions for which the average calorie supply falls considerably short of the corresponding requirement, indicating the existence of undernutrition. The proportion of undernourished, for the world as a whole, would thus seem to be between 10% and 15%. In other words, as a very conservative estimate, between 300-500 million people in the world are undernourished today.

It is interesting to use the foregoing approach to analyze the situation in the developed countries. Table 18 presents two calorie distributions on reference-man basis, one for rural households of the north-central region of the United States based on the food consumption survey conducted in 1952, and the other from a nationwide survey conducted in 1955. In a well-developed country one would expect most households to have sufficient calorie supplies. The calorie requirement at the retail level for the reference man in the United States is approximately 3,500 and the corresponding standard deviation per household on the basis of its average size of 2.5 reference men can be placed at approximately 300 Calories per reference man. One would therefore expect most households to have Calorie supplies exceeding 2,600 per reference man. The data shown in Table 18 accord with these expectations. In a well-fed country however one would expect that not only would there be no under-nutrition but that the average intake would equal the average require-ment, and further, that the variance of intake would also be equal to the variance of requirement.

TABLE 18. - DISTRIBUTION OF HOUSEHOLDS BY CALORIE SUPPLIES PER REFERENCE MAN PER DAY, UNITED STATES

Calories per reference man per day	Percentage frequency					
	North-central region U.S.A., 1952		U.S.A., 1955			
	Rural farm	Rural nonfarm	Rural farm	Rural nonfarm	Urban	Total
Under 2 000	2	3	5 (Under 2 000 to 2 500-3 000)	10 (Under 2 000 to 2 500-3 000)	14 (Under 2 000 to 2 500-3 000)	12 (Under 2 000 to 2 500-3 000)
2 000-2 500	3	6				
2 500-3 000	12	14				
3 000-3 500	19	22	9	10	14	12
3 500-4 000	22	19	11	17	17	16
4 000-5 000	27	25	28	29	27	28
5 000-6 000	15 (5 000 and over)	11 (5 000 and over)	20	17	15	16
6 000 and over			27	17	13	16

The data for the United States show that on the average the calorie supply purchased per reference man exceeds the requirement at the retail level by 500 and, further, as indicated in Table 18, that the number of households exceeding the requirement plus 3 times the standard deviation per household on a reference-man basis ap-proximates to 30% for the north-central region and indeed ap-proaches 50% for the nation as a whole. This could indicate that a large proportion of people in the States is probably overeating. It is of course possible that intake may have been overestimated, due, among other factors, to the underestimation of wastage. Alter-natively, requirement could have been underestimated. Possibly all three situations exist.

Malnutrition

Evidence of malnutrition

Malnutrition reflects the inadequacy in the nutritional quality of the diet. Diets of poor nutritional quality are common in most of the less developed areas, as indicated by the high percentage of calories derived from foods rich in carbohydrates and the small consumption of animal protein. This is due to excessive dependence on cereals and starchy foods and low consumption of animal foods. In addition, consumption of other protective foods such as fresh green and yellow vegetables and fruits is inadequate. These factors, together with wrong methods of food preparation and religious taboos and traditional prejudices are responsible for the deficiency of good quality protein and of essential minerals and vitamins. Diets of poor nutritional quality and/or insufficient quantity are largely responsible for listlessness, general impairment of health, poor physical development, and low resistance to infections and diseases. They are directly responsible for the widespread occurrence of specific deficiency diseases. They also contribute to the high mortality among infants and young children and the low expectation of life in the less developed areas.

The protein-calorie deficiency diseases (kwashiorkor and marasmus) result from the provision to weaned infants of diets which lack sufficient protein of good quality and which often do not provide enough energy for the efficient utilization of this limited amount of protein. They are the commonest of the nutritional deficiency diseases in the world today. A nutritional survey in Uganda, where plantains are particularly important in the diet, has revealed an incidence of kwashiorkor ranging from 6 to 11% in different groups of young children examined. Surveys in southern Nigeria, where yams and cassava are the staple foods, have indicated an incidence of kwashiorkor among young children of about 5%, significantly higher than the incidence of 2% found in the northern region of Nigeria, where millets are the staple foods. An extraordinarily high incidence of kwashiorkor characterizes the Feshi district of the southern Congo (Leopoldville), where cassava provides more than 80%

of the total calorie intake. The incidence of these protein-calorie deficiency diseases is also high in Latin America and the Far East. In general it coincides with the consumption of diets in which maize and cassava are the staple foods, and is related to the lack of milk and other good quality protein foods in children's diets especially in the postweaning period. The reported incidence refers to cases with well-defined clinical signs. Less serious cases of protein-calorie malnutrition are much more common. It has been claimed that most Africans have suffered from the disease at some time in their childhood often with permanent aftereffects. An increasing use of pulses, as in India and Pakistan, and of milk and other animal products, as in Japan and China, Taiwan, is helping to reduce the incidence of protein malnutrition.

Vitamin A deficiency is mainly due to the insufficient intake of foods like green and yellow vegetables, milk, eggs, butter, fishliver oils, and carotene-containing vegetable oils such as red palm oil. It is still widespread in many countries of the Far East, Latin America, and Africa. The importance of vitamin A deficiency is that total and incurable blindness often results, particularly among children. It also causes night blindness and skin disorders. Vitamin A deficiency is a public health problem in Indonesia, Mainland China, Burma, and elsewhere in the Far East, in parts of Latin America and in the semiarid zones of Africa. Throughout the humid coastal zones of west Africa supplies of carotene (pro-vitamin A) are ample because of the widespread use of red palm oil.

Nutritional anaemia which may result from dietary deficiencies of iron, folic acid, or other nutrients accentuated by intakes of protein poor in quantity or quality, is common in areas where starchy roots are the staple foods and green leafy vegetables are eaten in only small amounts. The condition is often aggravated by chronic parasitic infestation which occurs among a high percentage of the population. Nutritional anaemia affects mostly pregnant and lactating women.

Owing to a low intake of milk and milk products the diets consumed in many parts of the world provide much less calcium than those eaten in the more developed countries. In certain developing areas where low levels of dietary calcium are associated with lack of exposure to sunlight, due either to climatic or cultural reasons, the incidence of rickets in children or bony deformities in their mothers may be high. This has been found in parts of India and Burma, in the Near East, and in north Africa.

Among the common signs of malnutrition observed in developing countries are sore lips and sore tongues, due to lack of vitamin B_2 (riboflavin). These result from the use of starchy staple foods which contain only traces of the vitamin and a low consumption of dairy produce (milk and eggs) and fresh leafy vegetables.

Endemic goiter is found in areas where the soil and drinking water are low in iodine, and is particularly prevalent among the inhabitants of remote granite mountains, but may also be found in certain low-lying areas. In these circumstances the provision and use of iodized salt is the best means of eradicating the disease.

Considerable progress has been made during the past decade in reducing the incidence of the deficiency diseases through the adoption by governments of measures planned to orientate food production, processing and utilization to the consumption of nutritionally satisfactory diets, the adoption of planned development of health improvement measures, the spread of nutrition and home economics education, and programs for better nutrition, especially for vulnerable groups. For example, beriberi which is due to the lack of thiamine virtually disappeared from areas in Indonesia, China (Taiwan), and India through the use of undermilled or parboiled rice. It is reported that undermilling of rice and the practice of not washing it before cooking has also been introduced in Mainland China. Government insistence upon the use of parboiled and undermilled rice has prevented the appearance of beriberi in Nigeria where rice production is expanding rapidly. On the other hand, the increasing use of machine milling and consumption of highly polished rice in some parts of the Far East has spread the disease. It is reported that beriberi is now common in Thailand, Burma, and the Republic of Viet-Nam, and also in south Mainland China, East Pakistan, and some parts of India. In parts of Japan, however, the thiamine deficiency which had resulted from the return to the prewar habit of consuming highly polished rice has been overcome by legislation for enrichment of rice with thiamine.

Indirect evidence of malnutrition is provided by vital statistics and records of growth rates of children. High infant mortality rates and poor growth rates in the less developed areas are indication of nutritional deficiency. Whereas the death rate for children under one year of age is less than 40 per thousand in industrial countries, it is about 100 per thousand in many Asian and Latin American countries and over 200 per thousand in some countries, particularly in Africa. Again, while the death rate of children between one and four years is some 1 per thousand in the United States and other developed countries, it exceeds 20 per thousand in many less developed countries.

Statistical data show that during the last 50 years a tremendous shift has taken place in the dietary habits of many of the well-developed countries towards animal products compared with that of cereals. By contrast, the intake of animal products including milk in many less developed countries, particularly in the Far East, even today is much below the level in the highly developed countries some 50 or 100 years ago. These differences in diet are significant in the light of the improvements in health which have taken place in these countries during the last 50 years or so. For example, expectation of life in France has increased from 54 in 1910 to 70 in 1956, while the corresponding improvement, say in India, though impressive, still does not bring expectation of life there to more than 40 years. Even if the greater part of the improvement is considered due to general advances in medicine and sanitary control, there is a general agreement that the improvements in the composition of the diet in most developed countries over the last few decades have contributed materially to the decrease in mortality rates among children, increase of longevity, and greater resistance to diseases. Further evidence is provided by the striking improvement in the stature and body weight of the people of Japan since the war and the simultaneous improvement in the quality of their diets through larger intake of protective foods.

Incidence of malnutrition

It will be clear already that there exists a large mass of evidence which shows that there is serious and widespread malnutrition in the less developed countries. Sufficient data obtained from clinical surveys are not available from which to estimate the incidence of specific deficiency diseases. Even if available, such data may not reveal sub-optimum states of nutrition. The incidence of malnutrition can therefore only be measured indirectly. In the absence of satisfactory information on individual nutrient intakes and requirements, the most that can be done is to estimate the incidence of malnutrition by reference to the nutritional quality of diets in the developed countries using an appropriate indicator for this purpose. Clearly a single indicator cannot be expected to reflect fully the many difficult factors which result in malnutrition; but indicators are often used to advantage in even more complicated subjects. For instance, a high national income per caput is reasonably taken to imply a high level of living but does not necessarily reflect an adequate level of living in every respect.

The indicator of quality of diet chosen in the second world food survey was the percentage of calories derived from cereals, starchy roots, and sugar and was also recommended by the United Nations and its Specialized Agencies as an indicator of the food consumption and nutrition component of the level of living. It is a simple indicator and will be used here. Clearly the choice of this indicator affects the estimate of the proportion of malnourished. It may understate malnutrition where starchy roots are preponderant and overstate it where cereals make a larger contribution to the calorie supply. Also some foods like white fish and poultry make a greater contribution to the animal protein supply than their contribution to the calories would suggest. So that in countries where a major contribution to animal protein supply is made by such fish, as in Japan, this indicator may well overstate malnutrition. Again, fresh fruits and vegetables which supply vitamins and minerals make a negligible contribution of calories, even when consumed in satisfactory amounts.

The proportion that is considered to be malnourished also depends on the standards chosen. Whatever reasonable standard is set, certain individuals who are on the wrong side of this standard will nevertheless succeed in providing themselves with an adequate diet by a careful selection of their foods. Even when the observed value of the indicator is high, stringent public health measures and nutrition education may be sufficient to avoid some of the worst consequences of malnutrition. But as the indicator rises, a stage is clearly reached when the knowledge, control, and education required to avoid all the consequences of malnutrition are so great that the only practical policy is to raise the general quality of the diet. On the other hand even in a country where the value of the indicator is low (like the United Kingdom) government measures are often considered necessary to avoid shortages of some specific nutrients. But the indicator is useful nevertheless because as it falls, the probability of malnutrition falls, and as it rises, the difficulty in avoiding malnutrition rises.

In the second world food survey whenever the proportion of calories from cereals, starchy roots, and sugar exceeded two thirds, this was taken as evidence of malnutrition, although no attempt was made to estimate the percentage of malnourished. To do this, the value of the indicator for the individual members of the population must be known. In the United Kingdom the average value of the indicator is 50% and the standard deviation is a little over 10%, so that in at least 90% of the households that indicator falls below 80%.

Table 19 shows a distribution of the indicator for all types of households in Maharashtra State (India). It will be seen that some 60% of the households derive more than 80% of their calories from cereals, starchy roots, and sugar. In other words, if this is representative of conditions in India as a whole, some 60% of the households live on diets substantially inferior in quality to those in the United King-

TABLE 19. - DISTRIBUTION OF HOUSEHOLDS BY THE PERCENTAGE OF CALORIES DERIVED FROM CEREALS, STARCHY ROOTS, AND SUGAR IN MAHARASHTRA STATE (INDIA) (1958)

Percentage of calories derived from cereals, starchy roots, and sugar	Percentage frequency
< 50	5.8
50 -	2.2
55 -	1.7
60 -	4.1
65 -	6.5
70 -	10.3
75 -	11.6
80 -	18.0
85 -	19.6
90 -	15.9
95 - 100	4.3
	100.0

dom and may be considered to be malnourished.[5] Data concerning the exact form of the distribution of the indicator throughout the world are very scarce; but distributions were fitted using the limited knowledge available concerning them and the average values for the various countries (see also Appendix 4). The conclusion was reached that the data on malnutrition for India were fairly typical of other less developed countries. So the incidence of malnutrition in the less developed regions is estimated at 60%.

To sum up, as a very conservative estimate some 20% of the people in the underdeveloped areas are undernourished and 60% are malnourished. Experience shows that the majority of the undernourished are also malnourished. It is believed therefore that relative to the nutritional levels enjoyed by the people in the well-to-do countries and relative to the actual calorie requirements of the countries based on the FAO international scale, some 60% of the people in the underdeveloped areas comprising some two thirds of the world's population suffer from undernutrition or malnutrition or both. Since there are undoubtedly some people in the developed countries who are also ill-fed it is concluded that up to a half of the peoples of the world are hungry or malnourished.

---

[5] For details and source of data, see Sukhatme, P. V., "The food and nutrition situation in India, Part I," *Indian Journal of Agricultural Economics*, Vol. XVII, No. 2, 1-28, 1962.

5. NUTRITIONAL TARGETS

Throughout the underdeveloped countries hunger and malnutrition are urgent problems and demand an immediate increase in the supply of food. The nutritional targets which form the central theme of this survey are therefore of the immediate type which people in the developing countries have a right to aspire to, overnight as it were, and have for their objective the elimination of undernutrition and a reasonable improvement in the nutritional quality of the diet. Such targets, hereafter called short-term targets, must (in the words of the Hot Springs Conference) be " based upon the practical possibilities of improving the food supplies of the population. " They have been drawn up to assist the developing countries in preparing plans in response to the United Nations " Development Decade " which calls upon the countries of the world to co-operate in achieving a rate of growth of at least 5% per annum in the gross national product. This rate of increase is necessary in order to cover the increase in the population and at the same time to provide for a reasonable increase in the level of living. In this context the short-term nutritional targets become possible by the year 1975. But by their very nature, targets set out what ought to be achieved, not necessarily what is likely to be achieved during a period of economic growth, and need to be implemented deliberately by appropriate planning and policy measures.

The short-term targets are only a first step in what must be a continuous effort to achieve improvement in health and physique through better nutrition. Nutritional goals are dynamic. [1] They will probably change as people change their way of life; they will certainly

[1] Wright, N. C., " The current food supply situation and present trends " in Russel, E. J. and Wright, N. C., ed., *Hunger: can it be averted?* London, 1961, British Association for the Advancement of Science.

change as the science of nutrition makes progress; and they will actually recede the more nearly they are realized, for increased stature and body weight lead to increased requirements for food. In this context long-term targets have been formulated, not with the intention of providing any final answer to the problems of hunger and malnutrition but to give an idea of the scope for improvement in the diet as the problems of poverty and scarcity become less acute. We consider the implications of trying to realize these objectives by the turn of the century.

Quantitative formulation of nutritional targets

The guiding principles in formulating nutritional targets are clearly (1) to ensure quantitative adequacy of diets, and (2) to improve the nutritional quality of diets.

1. The objective of eliminating undernutrition requires that the calorie target should be equal to the corresponding average calorie requirement at the physiological level plus whatever allowance is needed to account for wastage and losses between the physiological level and the retail level (i.e., the level at which food supplies are usually measured). Further, the calorie supply should be so distributed that everyone in the population eats enough to meet his energy needs. In practice however the calorie supply will be unequally distributed, with some people taking more calories than they need. The calorie targets at the retail level need therefore to be so set up that they are equal to the average requirement at the physiological level increased by an appropriate amount to allow for wastage between the retail and physiological levels, and also to cover inequalities in distribution between social strata and within families.

The losses between the retail level (as brought into the kitchen) and the physiological level include losses of foods through spoilage, losses in cooking, wastage on plates, and food fed to domestic animals. However, losses through spoilage in the underdeveloped countries occur mostly during storage in homes in rural areas, owing to poor storage facilities. They are admittedly larger than in the developed countries. But these losses occur all through the year before the food is brought into the kitchen. They should strictly be taken into account in estimating the available food supplies at the retail level and therefore do not enter into the difference between the retail level and the physiological level. On the other hand, losses in the preparation of foods and wastage on plates are much smaller than in the developed countries. The quantity fed to domestic animals in developing countries is difficult to estimate. Altogether, it would appear that the losses between the retail and physiological levels in the developing countries are lower than in the prosperous countries and for the purposes of this survey have been placed at 5-10%. As the proportion of the rich to poor is exceedingly small in most developing countries, the over-all allowance of 10-12% over and above the calorie requirement at the physiological level leaves enough room for extra consumption arising from inequalities in distribution between social strata and within families. Accordingly, the *short-term targets* for the developing regions are set at the current requirement at the physiological level increased by 10-12%.

The dynamic nature of nutritional targets already referred to demands some extra allowance for improved stature and health. The *long-term calorie targets* for developing regions are set higher by 50-100 Calories to allow for an increase in height and body weight of adults. The corresponding increases for children are allowed for in the short-term targets.[2]

In the developed regions, as shown in sections 3 and 4, there is almost always an excess of calorie intake over requirements. This can only be explained either by increasing stature, underestimation of physical activity, or underestimation of wastage of food between the retail and physiological levels. There is not adequate evidence to conclude that people in the developed countries are overeating to anything like the extent indicated by the gap. Part of the obesity observed might well be due to lower activity than needed for a healthy active life. It follows that a good part of the gap must be explained by either underestimation of calorie requirements or

[2] The physiological energy requirements up to the age of 15 years suggested by the Second FAO Committee on Calorie Requirements are the same (except for an allowance of environmental temperature) for children in developing and developed countries.

of wastage between the retail and physiological levels. The existence of obesity suggests the desirability of reducing the current consumption levels. This is however hardly practicable in the short run, and even in the long run the effects of increased income on consumption and wastage may counterbalance the effect of any inducement to reduce the calorie intake, however desirable nutritionally. Accordingly, no change is made in the current consumption levels for the developed regions.

2. The objective of improving nutritional quality of diet requires that the intake of protective foods should be increased to a level approaching that in the well-fed countries, insofar as this is feasible and advisable. This objective cannot be achieved merely by specifying targets for the total amount of proteins in the diet. Proteins refer to only one aspect of the nutritional quality of diets and the quality of proteins from different sources varies. However, most diets rich in good quality proteins like those of animal origin also include a greater variety of foods and are therefore likely to contain substantial amounts of other essential nutrients. It has accordingly been a custom to use protein derived from animal sources as a measure of the quality of the diet, provided that the total protein is adequate.

To achieve optimal utilization of protein for growth and maintenance of health, it is essential that the diet provide adequate energy from nonprotein sources (fats and carbohydrates). While fulfilling its primary function of providing for growth and maintenance, dietary protein contributes 4 Calories for each gram so utilized. If the diet does not supply sufficient energy from nonprotein sources, protein is diverted from its primary functions of providing for growth and maintenance of body tissues to supplying energy for other vital functions. Provided adequate energy is available from fats and carbohydrates, the total amount of protein required for growth and maintenance depends inversely upon its quality (biological value). In general the proteins derived from animal sources have a higher biological value than those derived from vegetable sources. Thus the greater the percentage of the over-all dietary protein which is derived from animal sources the smaller will be the requirement and the fewer will be the calories derived from its utilization, always provided the supply of non-protein calories is adequate. Provided that a reasonable proportion of total protein is derived from animal foodstuffs, a diet is likely to supply sufficient protein if 10-12% of the calories are derived from this nutrient.

In well-fed countries like the United Kingdom and France, about half the dietary protein is of animal origin. To aim at so high a proportion as one half or even a third for the developing countries is clearly not practicable, for animal proteins are far more costly than vegetable proteins; and even if it were possible to produce them most people would not be able to afford them until their purchasing power had gone up proportionately. But such high proportions may be unnecessary. Protein malnutrition is mainly important in infants and children. Protein deficiency in adults and adolescents (except pregnant and nursing mothers, the ill and infirm) can be remedied by adequate amounts of mixtures of vegetable foods; for although individual vegetable products are known to be deficient in one or more of the essential amino acids, a mixture is usually not.

Taking the above considerations into account, the *short-term targets* for the developing regions provide for a modest increase of 5-6 grams of animal protein to a total of 15 grams per person per day, which amounts to 22% of the total proteins. If distributed according to physiological needs within populations and families, this would go a long way to meet the most urgent needs of the vulnerable groups. A larger increase would have been desirable, but so small has been the trend of increase that the attainment of a larger target in the course of a decade or so is clearly outside the range of practical feasibility. Neither would a larger target be within the purchasing capacity of the people, even if Member Governments were to succeed in increasing the gross national product at a rate approaching 5% per annum in response to the call of the United Nations Development Decade.

The *long-term target* for the developing regions has been increased by a further 6 to a total of 21 grams per person per day which is equivalent to 28% of the total protein. This is intended to cover more adequately the needs of the vulnerable groups and to satisfy in some measure the demand in other sections of the community.

In the developed countries the intake of animal and total protein 56

TABLE 20. - SHORT-TERM AND LONG-TERM TARGETS FOR CALORIES, TOTAL PROTEINS, AND ANIMAL PROTEINS, BY REGIONS (PER CAPUT PER DAY)

Regions	SHORT-TERM TARGET					LONG-TERM TARGET				
	Calories	Total protein	Animal protein	Percentage of animal protein of total protein	Percentage of total calories derived from protein	Calories	Total protein	Animal protein	Percentage of animal protein of total protein	Percentage of total calories derived from protein
		...... Grams					...... Grams......			
Far East	2 300	68	12.5	18	12	2 400	74	20	27	12
Near East	2 450	77	20	26	12	2 500	79	25	32	12
Africa	2 400	69	18	26	12	2 500	75	25	33	12
Latin America (excl. River Plate countries)	2 550	71	25	35	11	2 550	71	25	35	11
Low-calorie countries	2 350	69	15	22	12	2 450	74	21	28	12
World	2 550	75	23	31	12	2 600	79	28	35	12

is often unnecessarily high, but in the face of consumer preferences the targets do not suggest any change from the present consumption levels.

The different targets for calories and animal and total proteins, region by region, are given in Table 20, both for the short-term target and the long-term target. Appendix 5 gives targets by subregions.

Limitation of targets

It should be re-emphasized that to improve nutrition in any given area, more is needed than an increase in the average level of calories, proteins, or other nutrients to the level implied in the targets. Satisfactory distribution of food within countries and families so that each may eat according to his needs is of utmost importance. Targets to be realistic ought therefore to be defined in relation to existing dietary patterns and preferences in the socio-economic groups within each country. In order to get the best results out of the supplies made available under the targets, it is necessary to direct them towards those members of the population who need them most, both through schools, clinics, hospitals, etc., and by means of associated measures of nutrition education. Such measures are best developed by the countries themselves, in the light of first-hand knowledge of the patterns of consumption and resources available in the countries. The objective of the third world food survey is only to give an idea of the order of change in food supplies needed to raise the peoples' level of nutrition. This objective is served by specifying targets for regions and subregions. It does not mean that a uniform degree of nutritional adequacy in all countries within a subregion is suggested. Indeed, in using subregional targets, differences between and within countries will be concealed.

6. FOOD SUPPLY TARGETS

How they are set up

The next step is to compute targets in terms of the major food groups which reflect the proposed nutritional goals and in addition provide for adequate amounts of nutrients other than proteins, that is vitamins and minerals. Two types of considerations were taken into account in establishing them. It was considered that no radical changes should be introduced into the existing dietary pattern of the people and that where they are called for, as in the case of diets in subregions of Africa and Latin America, they would be introduced gradually as one passes from the short-term target to the long-term. The second consideration, which is rather more relevant for the short-term target (but will probably always be relevant to some extent), is economic and production feasibility, and ensures that the computed targets are feasible of achievement and as far as the short-term goals are concerned are within reach of the people in the course of the next decade or so. The long-term production possibilities can only be speculated upon, but every possible care has been taken in formulating the long-term supply targets.

The targets are expressed at the retail level. The sum of the calories derived from cereals, starchy roots, and sugar has been substantially reduced to allow for the inclusion of satisfactory amounts of pulses, animal products, fats, and oils in the diets. The specific points for each food group taken into account in establishing the food supply targets within the over-all considerations given above are set out below.

Cereals

In most of the developing regions and subregions this food group makes a greater contribution than any other food group, not only to the energy value of the diets but also to their protein content.

In some subregions, notably central Africa, starchy roots contribute as much to the calorie supply as cereals, but even here cereals provide more protein than any other single food group. The biological value of many cereal proteins is high, particularly those of rice and certain millets, and they provide a more concentrated form of energy than root crops. Therefore, in formulating the targets, cereals have been used to replace a part of the starchy roots in areas where consumption of the latter is excessively high. In other developing regions targets for cereals are reduced to allow for the additional consumption of foodstuffs from animal sources needed to meet the animal protein targets. In general, the targets approximate to, or are set below the current levels.

Starchy roots

In some areas these staple foods can be produced very easily and provide the cheapest source of calories. Consequently the calorie target for such subregions could be met most economically by increases in supplies of roots like cassava and plaintains alone. This however is not nutritionally desirable because the calorie concentration in such foods is low and the physical bulk of these diets is very great, an undesirable nutritional characteristic which is of great importance in the case of small children and expectant mothers. Moreover, the protein content is low and of poor quality. The food consumption targets for subregions in which the energy value of the diets consumed is largely derived from starchy roots have been so determined that a proportion of the energy presently derived from roots is replaced by more concentrated sources of calories such as cereals, fats, and oils.

Sugar

In certain developing subregions it would appear that sugar consumption is somewhat too high, and the money spent on it might be better used for the purchase of more nutritious food. Even though sugars may be a reasonably cheap source of energy in these areas, and the yield per acre higher than for cereals, most of them are purely carbohydrates. Therefore, for such subregions the target has been set below the present consumption. On the other hand, in other developing subregions the present levels of sugar consumption would appear low, sometimes contributing less than 50 Calories to the total, due among other things to the relatively high cost of sugar compared with other carbohydrates. Nevertheless, recent estimates of income elasticity of demand for sugar are high in these areas, suggesting the likelihood of considerably increased consumption with increasing per caput income. The increase over present consumption suggested in the targets has been placed for nutritional reasons below the level expected on economic grounds.

Pulses and nuts

This food group is nutritionally important as a source of energy as well as protein, some minerals, and B-complex vitamins. In regions where the present consumption of proteins is low and the potential production of foods from animal sources is poor, the target for pulses and nuts has been set at a relatively high figure.

Fruit and vegetables

A considerable increase in the production and consumption of fresh fruit and green leafy vegetables is necessary in many of the developing areas of the world in order to provide many of the minerals and vitamins essential for good nutrition. In particular, adequate provision of this food group in the diet is important for the prevention of specific deficiency diseases such as scurvy, beriberi, and keratomalacia, and for satisfactory growth and maintenance of health. On the other hand, the production of many fruits and vegetables in semiarid, arid, and cold climates is not easy and contributes greatly to the cost of the diet. In line with the first world food survey, the principle followed in establishing targets for this food group was that it should contribute not less than 5% of the calories to the total energy value except in some subregions of the Far East and Africa where the current level is abnormally low and any increase to bring the current level of supply to 5% of the calorie needs is not considered feasible in the short-term target. Adequate provision, however, is made in the long-term targets.

Products of animal origin (meat, fish, eggs, milk and milk products, excluding butter)

Criteria for setting up the animal protein targets have been described in the previous section. Of the different animal products, eggs and meat are usually the most costly sources of proteins and also of calories. Considerations of palatability and the desirability of ensuring a steady supply of animal proteins with the associated minerals and vitamins throughout the year made it necessary, however, to ensure that a certain minimum amount of animal proteins from these sources was included in the diet by establishing targets higher than the existing levels but consistent with what is agriculturally feasible and what the consumer can afford by 1975. Where it is available, fish is usually the cheapest source of animal protein. The target for animal protein could therefore be most economically met by increases in fish supplies. On nutritional grounds there is no objection to fish as a sole source of animal protein. An upper limit was set for the fish supply targets to take account of production feasibility and consumer preferences. Milk and milk products are not always cheap sources of animal protein, but are valuable sources of calcium. Milk is the most convenient food for meeting most of the nutritional requirements of infants, children, expectant and nursing mothers, the old, and the sick. The suggested target increases largely reflect the needs of these vulnerable groups.

Fats and oils

Fats and oils obtained from foodstuffs of animal and vegetable origin are a concentrated source of energy, and certain of them supply large amounts of fat-soluble vitamins and essential fatty acids.

Provided a diet supplies enough energy and sufficient of these essential food factors, nutritionists have not yet determined a minimum requirement for fat *per se*. On the other hand, diets which contain large amounts of fat (up to 40% calories from fat) may be harmful to health. But this consideration does not apply to the low-calorie countries where the inclusion of reasonable quantities of fat, particularly of unhydrogenated vegetable oils (good sources of essential fatty acids) in a diet which would otherwise rely on starches as the main source of energy, increases the calorie value of the diet while reducing its over-all bulk. This is an important factor in child feeding, particularly in developing countries. On a subregional basis such dietary improvements cannot be reflected fully in the figures. Still, in most subregions the target for fats and oils calls for a substantial increase in the short-term targets. The long-term targets provide for approximately double the present consumption in the low-calorie countries. Larger increases would have been nutritionally desirable but are hardly feasible when production and economic considerations are taken into account. These increases in visible fats have been accompanied by corresponding increases in the invisible fats, especially those contained in milk and meat.

Food supply targets in comparison with present consumption levels

Tables 21 to 26 show region by region the data on per caput food supplies available and needed under the short-term and long-term goals. The tables also show the resulting total values for calories, total proteins, animal proteins, and other indicators of the nutritional quality of diet, together with price-weighted index numbers of per caput food supplies [1] required to attain the respective targets. Appendix 6 shows the details by subregions.

A comparison of the food supplies available and needed to attain the short-term target shows that the Far East with a population of over 1,500 millions in 1958 has a current deficit in the protective

[1] See note 1 to Table 8.

food groups of some 15 million tons in pulses, 45 million tons in fruit and vegetables, 40 million tons in animal products, 4 million tons in fats and oils. Over-all food supplies today provide only some three fourths of what is essential for a healthy, active life. The deficit is particularly large in animal products which provide only about

TABLE 21. - PER CAPUT FOOD SUPPLIES AVAILABLE AND NEEDED UNDER THE SHORT-TERM AND LONG-TERM TARGETS TOGETHER WITH THE CALORIE AND PROTEIN LEVELS, FAR EAST

Item	Available	Short-term target	Needed/ available %	Long-term target	Needed/ available %
	...... *g/day*			.. *g/day* ..	
Cereals	400	395	99	361	90
Starchy roots	166	160	96	144	87
Sugar	24	35	146	35	146
Pulses and nuts	50	75	150	80	160
Vegetables and fruit	144	225	156	315	219
Meat	24	36	150	66	275
Eggs	3	6	200	8	267
Fish	12	21	175	30	250
Milk	54	98	181	140	259
Fats and oils	9	16	178	24	267
TOTAL CALORIES	2 060	2 310		2 400	
% calories derived from cereals, starchy roots, sugar	81	73		64	
TOTAL PROTEINS (g/day)	56	68		74	
ANIMAL PROTEINS (g/day)	7.5	12.5		20	
Over-all index of per caput food supply		131		170	
Over-all index of per caput animal food supply		163		267	

60% of what is nutritionally desirable under the short-term target. The deficits for the different subregions range from 15 to 30% in over-all food supplies and from 20 to 45% in animal products. The inadequacy of the current food supply is even more striking when compared with the long-term target. The protective foods, including fats and oils, are much less than a half of what is desirable as a long-term goal and the diet as a whole falls short of it by some 40%.

These deficits mean that the available food supplies per person must be increased by nearly one third and animal products by two

TABLE 22. - PER CAPUT FOOD SUPPLIES AVAILABLE AND NEEDED UNDER THE SHORT-TERM AND LONG-TERM TARGETS, TOGETHER WITH THE CALORIE AND PROTEIN LEVELS, NEAR EAST

Item	Available	Short-term target	Needed/ available %	Long-term target	Needed/ available %
	...... *g/day*			.. *g/day* ..	
Cereals	446	401	90	374	84
Starchy roots	44	44	100	44	100
Sugar	37	44	119	50	135
Pulses and nuts	47	47	100	47	100
Vegetables and fruit	397	397	100	397	100
Meat	35	53	151	68	194
Eggs	5	10	200	25	500
Fish	6	7.5	125	15	250
Milk	214	307	143	307	143
Fats and oils	20	25	125	30	150
TOTAL CALORIES	2 470	2 470		2 500	
% calories derived from cereals, starchy roots, sugar	72	66		62	
TOTAL PROTEINS (g/day)	76	77		79	
ANIMAL PROTEINS (g/day)	14	20		25	
Over-all index of per caput food supply		117		130	
Over-all index of per caput animal food supply		148		180	

thirds, in order to attain the short-term goal. The increases needed to attain the long-term target are much higher.

The index number of total food supply in Africa falls short of that needed under the short-term target by some 20% and that of animal products by some 40%. Compared to the long-term target the total food supply is two thirds and that of animal products considerably less than one half.

For the Near East, deficits are smaller, being 15% in total foods and some 30% in animal foods under the short-term target. The

TABLE 23. - PER CAPUT FOOD SUPPLIES AVAILABLE AND NEEDED UNDER THE SHORT-TERM AND LONG-TERM TARGETS, TOGETHER WITH THE CALORIE AND PROTEIN LEVELS, AFRICA

Item	Available	Short-term target	Needed/ available %	Long-term target	Needed/ available %
	...... *g/day*			.. *g/day* ..	
Cereals	330	365	111	340	103
Starchy roots	473	240	51	224	47
Sugar	29	31	107	31	107
Pulses and nuts	37	44	119	44	119
Vegetables and fruit	215	232	108	317	147
Meat	40	60	150	74	185
Eggs	4	10	250	15	375
Fish	8	20	250	35	438
Milk	96	145	151	203	211
Fats and oils	19	20	105	25	132
TOTAL CALORIES	2 360	2 400		2 500	
% calories derived from cereals, starchy roots, sugar	74	68		62	
TOTAL PROTEINS (g/day)	61	69		75	
ANIMAL PROTEINS (g/day)	11	18		25	
Over-all index of per caput food supply		123		151	
Over-all index of per caput animal food supply		171		239	

corresponding deficits under the long-term target are nearly 25% and 45%.

The food deficits are the least in the case of Latin America (excluding the River Plate countries), where per caput supplies approach more nearly those in the prosperous countries. Therefore, only one target has been set for this region. Total food supplies fall short of this target by a little over 10% and those of animal foods by a little over 20%.

For the low-calorie countries together the food supply per caput is only some 80% of what is urgently required, and 65% of what is

TABLE 24. - PER CAPUT FOOD SUPPLIES AVAILABLE AND NEEDED, TOGETHER WITH THE CALORIE AND PROTEIN LEVELS, LATIN AMERICA, EXCL. RIVER PLATE COUNTRIES

Items	Available	Target	Needed/ available %
	g/day		
Cereals	282	315	112
Starchy roots	225	169	75
Sugar	89	74	83
Pulses and nuts	53	53	100
Vegetables and fruit	355	355	100
Meat	72	85	118
Eggs	11	16	145
Fish	8	16	200
Milk	201	250	124
Fats and oils	22	25	114
TOTAL CALORIES	2 430	2 550	
% calories derived from cereals, starchy roots, sugar	65	62	
TOTAL PROTEINS (g/day)	63	71	
ANIMAL PROTEINS (g/day)	19	25	
Over-all index of per caput food supply		114	
Over-all index of per caput animal food supply		128	

needed under the long-term target. Per caput supplies of animal foodstuffs are 65% and 45% respectively under the two types of goals.

The net result for the world is that per caput food supplies are of the order of 90% of what is required under the short-term target and 80% under the long-term target. The corresponding figures for animal foods are of the order of 85% and 70%.

The results of the survey imply that for the world as a whole there is an immediate deficit of some 60 million tons in animal products,

TABLE 25. - PER CAPUT FOOD SUPPLIES AVAILABLE AND NEEDED UNDER THE SHORT-TERM AND LONG-TERM TARGETS TOGETHER WITH THE CALORIE AND PROTEIN LEVELS, LOW-CALORIE COUNTRIES

Item	Available	Short-term target	Needed/ available %	Long-term target	Needed/ available %
	...... g/day			.. g/day ..	
Cereals	386	386	100	356	92
Starchy roots	194	162	84	148	76
Sugar	31	39	126	39	126
Pulses and nuts	48	68	142	72	150
Vegetables and fruit	184	248	135	324	176
Meat	30	44	147	69	230
Eggs	4	8	200	10	250
Fish	12	20	167	28	233
Milk	80	129	161	166	208
Fats and oils	12	18	150	24	200
TOTAL CALORIES	2 150	2 350		2 430	
% calories derived from cereals, starchy roots, sugar	78	71		64	
TOTAL PROTEINS (g/day)	58	69		74	
ANIMAL PROTEINS (g/day)	9.5	15		21	
Over-all index of per caput food supply		127		157	
Over-all index of per caput animal food supply		157		233	

50 million tons in fruit and vegetables, and 5.5 million tons in fats and oils. The long-term target indicates the need for increasing current production by twice these deficits to meet the needs of the present world population, while requiring some 20 million tons less of cereal for human consumption.

TABLE 26. - PER CAPUT FOOD SUPPLIES AVAILABLE AND NEEDED UNDER THE SHORT-TERM AND LONG-TERM TARGETS TOGETHER WITH THE CALORIE AND PROTEIN LEVELS, WORLD

Item	Available	Short-term target	Needed/ available %	Long-term target	Needed/ available %
	...... *g/day*	......		.. *g/day* ..	
Cereals	367	367	100	347	95
Starchy roots	230	207	90	197	86
Sugar	49	53	108	53	108
Pulses and nuts	38	52	137	55	145
Vegetables and fruit	238	282	118	336	141
Meat	66	76	115	94	142
Eggs	11	15	136	16	145
Fish	14	19	136	25	179
Milk	225	261	116	287	128
Fats and oils	22	27	123	31	141
TOTAL CALORIES	2 420	2 560		2 620	
% calories derived from cereals, starchy roots, sugar	70	66		62	
TOTAL PROTEINS (g/day)	68	75		79	
ANIMAL PROTEINS (g/day)	20	23		28	
Over-all index of per caput food supply		111		123	
Over-all index of per caput animal food supply		116		138	

7. THE SIZE OF FUTURE FOOD NEEDS

The discussion in the previous section is confined to the dietary changes needed to achieve better levels of nutrition. It does not deal with the increases required to meet the needs of the growing population. The United Nations has prepared population projections up to the year 2000 based on various assumptions - low, medium, and high. Revised regional figures for 1975 and 2000 based on the medium assumption are shown in Table 27. Subregional figures are given in Appendix 7. These figures imply that increases in population alone without any improvement in the existing level of diet would call for an over-all increase in food supplies in the low-calorie countries of 41% by 1975 and 150% by the year 2000. The increases are much lower in the high-calorie countries. For the world as a whole food supplies would have to be increased by 36% by 1975 and 123% by 2000. Estimates of the index number of total food supplies covering the requirements for better levels of nutrition and for increasing population are shown in Table 28. The size of the problem involved in achieving the short-term targets (by 1975) and the long-term targets (by the year 2000) is discussed below. Appendix 8 gives details by regions and subregions.

Low-calorie countries

Far East

Interpreted in terms of future needs, the short-term target implies that for every 100 million increase in population, the region would have to provide a total of some 15 million tons of additional cereals, 3 million tons of additional pulses, 8 million tons of additional fruit and vegetables, and 6 million tons of additional animal products. According to the medium projection the population in the Far East is expected to grow from a little over 1500 millions in 1958 to some 2150 millions in 1975. It follows that if the short-term target is to be achieved by 1975 the total food supplies in the region would have to be increased by some 40% in cereals, 110% in pulses, 120% in fruit and vegetables, 155% in milk, and 150% in fats and oils (Appendix 8). Expressed in terms of the over-all index of animal food

TABLE 27. - PROJECTED GROWTH OF POPULATION AND INDEX NUMBER OF POPULATION, BY REGIONS (1958 = 100) (medium assumption)

Regions	1975		2000	
	Population in millions	Index number	Population in millions	Index number
Far East	2 150	140	3 753	245
Near East	182	143	326	257
Africa	273	133	458	223
Latin America (excl. River Plate countries)	272	155	552	315
Latin America, River Plate countries	32	128	43	172
Europe	757	121	954	153
North America	250	130	325	169
Oceania	22	146	30	200
Low-calorie countries ..	2 877	141	5 089	250
High-calorie countries ..	1 061	124	1 352	158
World	3 938	136	6 441	223

and food supplies as a whole, the increases needed would be 128% and 83% respectively. These represent an annual compound rate of increase of 4.9% in animal foods and 3.6% in foods as a whole.

The trend of increase in over-all food supplies over the last 25 years has barely kept pace with the population growth (Table 8). There is, however, an indication that in some countries of the Far East the trend has exceeded population growth and reached a rate of 3% during the last decade. Encouraging as this development is, this rate would have to be stepped up and maintained at well over 3.5% for the region as a whole in order to achieve the short-term target by 1975. In actual fact the available evidence does not indicate any increase in the per caput supply level between 1958 and 1962. If this is so, the actual rate needed to achieve the target by 1975 is more nearly 4.5% than 3.5% in total food supplies (this implies a rate of increase of 2.3% in per caput food supplies, the rate of population increase being 2%). Any attempt, however, to push up the rate of increase in per caput food supplies has to be accompanied by a corresponding effort to increase the purchasing power of the people so that they may absorb the additional food supplies available. It follows that if the aim is to attain the short-term targets by 1975 economic development plans should provide an increase in per caput income which, with the known income elasticity for food as a whole,

TABLE 28. - OVER-ALL INDEX NUMBER OF NEEDS IN TOTAL AND ANIMAL FOOD SUPPLIES UNDER THE SHORT-TERM AND LONG-TERM TARGETS, BY REGIONS (1957-59 = 100)

Regions	Short-term target 1975		Long-term target 2000	
	Total foods	Animal foods	Total foods	Animal foods
Far East	183	228	417	654
Near East	167	212	334	463
Africa	164	227	337	533
Latin America (excl. River Plate countries)	176	198	359	403
Low-calorie countries ..	179	221	393	583
World	151	158	274	308

would support a rate of increase in per caput food supplies of 2.3%. The income elasticity of food as a whole being of the order of .8, this rate of increase in per caput income is estimated at some 3% per annum. At a rate of population growth of approximately 2% per annum (and indications are that population growth is likely to be faster), the aim should therefore be to increase the aggregate national income by approximately 5% per annum, which in fact is the target for the decade 1960-70 in the plans for activating economic development through the joint efforts of the United Nations, FAO, and other specialized agencies.

The long-term target calls for larger increases in food supplies. Assuming that their realization is contemplated by the year 2000 and that the population were about 2.5 times its present size, total food supplies should be more than quadrupled, and the supplies of animal foodstuffs should be 6½ times their present amount.

Other developing regions

The implications of the targets have been discussed above in detail for the Far East, the largest of the developing regions. No detailed discussion for the other developing regions is made as the data presented in Tables 27 and 28 tell much the same story.

Low-calorie countries as a whole

Putting the results for the low-calorie countries together, we find that the total supplies in cereals would have to be increased by some 40% by 1975 and 130% by the year 2000; the corresponding figures are 100% and 275% for pulses and 120% and 485% for animal products. The respective increases in food as a whole needed by the low-calorie countries are 79% by 1975 and 293% by the year 2000.

The increase in total food supplies by 1975 calls for a per caput rate of increase of 2.0%, corresponding to some 3% in per caput income (the income elasticity for food as a whole being .7). If the population growth rate is 2% (and it may well be higher) this implies a rate of increase in aggregate national income approximating 5% which again is compatible with the philosophy of the United Nations "Development Decade."

The world

For the world as a whole, total food supplies in 1975 would have to increase by some 35% in cereals, 85% in pulses, and 60% in animal products, and in 2000 by some 110% in cereals, 225% in pulses, and 210% in animal products. The over-all change in the world's total food supplies needed to achieve the short-term target by 1975 and the long-term target by the year 2000 are 51% and 174% respectively. In actual fact, the necessary increase in food supplies may well be greater because of inequality of distribution between countries. So the broad conclusion is that should the world population grow according to the United Nations medium projection its total food supply would have to be increased by well over a half by 1975 and threefold by the turn of the century, in order to provide a satisfactory level of nutrition.

8. THE POSSIBILITIES OF PRODUCING THE REQUIRED FOOD

What are the possibilities of producing the needed foods? This topic is the theme of another Basic Study prepared by FAO for the Freedom from Hunger Campaign and will therefore be touched upon only briefly here. [1]

Production possibilities were kept in mind when setting up the short-term targets, but one is left wondering whether the large increases can in fact be achieved over a decade or so. These doubts may be even stronger in the case of the more distant objective. But there should be little room for doubt on one score: the world could grow enough food to meet all these needs if we made rational use of nature's bounty. Doubts arise because the necessary effort to reach the targets may not be made. Other objectives may cloud the horizon: the financial resources required to develop the world's agriculture may not be made available; the incentive to expand production may be absent in many parts of the developing world and the underdeveloped countries may be unable to purchase the food supplies they require.

If present technical knowledge can be spread, potential food production is extremely large. The use of organic and inorganic fertilizers, control of pests and disease, better seed and appropriate methods of cultivation, together with the large possibilities of extending irrigation and double-cropping, provide the means of increasing crop and fodder yields. Better and scientific feeding, timely use of forage crops, control of animal disease, and an efficient breeding program can likewise increase yields of livestock. Without any expansion of the cultivated area the production of crops could well be doubled. For livestock products the prospects are even better, although yields as

[1] *Possibilities of expanding world food production.* FFHC Basic Study No. 10, FAO, Rome, 1963.

high as those in some developed countries are not likely to be obtained until the scarcity of feeding stuffs has been overcome.

Lack of adequate soil surveys makes it difficult to say how much land is capable of producing crops, but the Basic Study referred to above shows that the area is very much larger than the 10% of the world's surface which is cultivated at present. Finally, a substantial increase in the effective yields can be made by improved methods of storage. For all these reasons it seems that supplies at the retail level could well be increased fourfold, and the true upper limit is probably much higher.

The agricultural resources can also be supplemented by a very great increase in the production of fish. The targets set imply an expansion of about fourfold by 2000, which may call for exploitation of marine resources outside the conventional fishing areas. There is also great scope for increasing production and improving the access to supply by deliberate culture of fish in inland lakes and waterways.

The position is more critical when we consider the requirements region by region. The low-calorie countries need to increase their food fourfold by 2000. In Latin America and Africa this expansion could be met by both increases in yields and expansion in the cultivated area, although this may not be achieved easily in areas of Africa where the custom of permanent settlement has scarcely been established. Where the shortage of land is acute as in the densely populated countries of the Far East and in the Near East, the problem is mainly one of raising yields through intensive agriculture. These countries must make the most of the "advantages of backwardness" quickly; otherwise the production may have little hope of keeping pace with the population.

In the high-calorie countries of the world, production seems more than likely to keep pace with the population growth. Even without any increase in incentive, a rise in the productivity of land of the order of 1½-2% is not merely possible but probable; and in addition in some areas (eastern Europe and the U.S.S.R.) a large acreage is available for new cultivation.

Clearly it is possible for countries with food surpluses to export food supplies to the low-calorie countries; but it is hardly conceivable that it can become a permanent means of redressing the imbalance between the regions, unless a final answer can be found to problems associated with the balance of payments. For example, in the case of the Near East where it is a little difficult to see how the extra food is going to be produced, the exports of oil may be sufficient to pay for food. But in general the importation of food by developing countries should take second place to increased national food production, and justifiably so. Surpluses, particularly of milk products and cereals, occurring in the rest of the world can make a contribution to the needs of the underdeveloped countries but only if very deliberate steps are taken to use them to increase their production potential. The World Food Program launched jointly by the United Nations and FAO on an experimental scale is a welcome indication of our awareness of this task.

APPENDIXES

The appendixes give detailed data by regions and subregions. The regions used in Appendixes 1 to 3 are defined in the FAO *Production yearbook* and *Trade yearbook*. The coverage of the regions and subregions used in Appendixes 4 to 8 (and, for reasons of consistency, of the subregions in Appendix 1) depends on the availability of data on food supplies and is given below:

FAR EAST

South Asia:	Ceylon, India, Pakistan
Southeastern Asia, mainland:	Burma, Fed. of Malaya, South Viet-Nam, Thailand
Eastern Asia:	China (Taiwan), Japan, South Korea
Southeastern Asia, major islands:	Indonesia, Philippines
China, Mainland	

NEAR EAST:	Cyprus, U.A.R., Iran, Iraq, Israel, Jordan, Lebanon, Libya, Sudan, Syria, Turkey

AFRICA

North Africa:	Algeria, Morocco, Tunisia
West and central Africa:	Congo (Leopoldville), Cameroon, former French Equatorial Africa, former French West Africa, Ghana, Guinea, Liberia, Nigeria, Togo
East and southern Africa:	Angola, Ethiopia, Kenya, Mada-

	gascar, Mauritius, Fed. of Rhodesia and Nyasaland, Tanganyika, South Africa
LATIN AMERICA	
Mexico and Central America:	Costa Rica, Cuba, Dominican Republic, El Salvador, Guatemala, Haiti, Honduras, Nicaragua, Mexico, Panama
Northern and western countries of South America:	Bolivia, Chile, Colombia, Ecuador, Peru, Venezuela
Brazil	
River Plate countries:	Argentina, Paraguay, Uruguay
EUROPE	
Western Europe:	Austria, Belgium, Denmark, Finland, France, Fed. Rep. of Germany, Greece, Ireland, Italy, Netherlands, Norway, Portugal, Spain, Sweden, Switzerland, United Kingdom, Yugoslavia
Eastern Europe:	Bulgaria, Czechoslovakia, E. Germany, Hungary, Poland, Romania
NORTH AMERICA:	Canada, United States
OCEANIA:	Australia, New Zealand
LOW-CALORIE COUNTRIES:	Far East, Near East, Africa, Latin America (excl. River Plate countries)

HIGH-CALORIE COUNTRIES: River Plate countries, Europe, North America, Oceania

Appendix 1 presents midyear estimates of the total population for 1938, 1950, and 1960 by subregions and regions. Wherever feasible, official estimates are shown; the unofficial estimates are mainly based on data furnished by the Statistical Office of the United Nations.

Figures given for 1938 have been revised in the light of recent information. In some cases unofficial interpolated values replace the official estimates that have not yet been revised to agree with the latest censuses.

The accuracy of the data is highly variable. In several areas, no comprehensive and reasonably reliable population count has been made recently and the estimates given for them are largely conjectural. In view of the fact that for several areas of the world the size of the population is unknown or uncertain, the regional and world totals must be regarded as approximate estimates only.

As far as possible the estimates relate to present boundaries and to the population present in the area, plus members of the armed forces stationed outside the area. Wherever sufficient information was available on prisoners of war, displaced persons, indigenous groups, or other groups ordinarily excluded from the official data of a country, the estimates have been adjusted to include them.

Appendixes 2 and 3 present index numbers of food production and per caput food production by regions for periods from prewar to 1961/62. Food production in these appendixes includes the following commodity groups: grains, starchy roots, sugar, pulses, oil crops, nuts, fruit, vegetables, wine, livestock, and livestock products. To avoid double counting, allowances are made for commodities used for livestock feeding; these include products fed as such, and semiprocessed feeds such as oilcakes and bran. In addition, to the extent that adequate estimates are possible, allowances are also made for imported feeds, seed, and production waste. The index numbers were constructed by applying regional weights, based on 1952-56 farm price relationships, to the production figures.

The data on per caput food supplies presented by subregions in Appendix 4 are based on national food balance sheets, partly prepared on a regular basis by governments in collaboration with FAO (and, in the case of European countries, with the Organization for Economic Cooperation and Development), and partly prepared as *ad hoc* estimates by FAO for the purposes of its world food surveys. The food balance sheet is a statistical method which starts from the data of production, trade, and movement in stocks for each foodstuff; makes appropriate deductions for the amounts used for feed, seed, manufacture, and waste; and so arrives at the estimated quantity of food supplied at the retail level and its nutritive value for a given period, usually a year. Food balance sheet data are subject to the following principal limitations: the range and accuracy of national statistics vary widely from country to country for a number of commodities, official statistics of production and trade are frequently inadequate, so that rough estimates have to be made. Except in a few countries, practically no data are available on farm and commercial stocks whose yearly changes may appreciably affect the supplies of food. Quantities utilized for feed, seed, waste and for industry are frequently roughly determined in the absence of statistics on the utilization of individual commodities. For these reasons, the results obtained for different countries are not fully comparable.

Data on population, index numbers of food production and on food supplies by countries are regularly published by FAO in its *Production yearbook.*

Appendix 1

POPULATION 1938-60 (MIDYEAR ESTIMATES) AND INDEX NUMBERS OF POPULATION (1938 = 100), BY SUBREGIONS AND REGIONS

Subregions and regions	1938	1950		1960	
	Population in 1 000	*Population in 1 000*	*Index numbers*	*Population in 1 000*	*Index numbers*
FAR EAST					
South Asia	380 135	441 011	116	533 743	140
Southeastern Asia, mainland	42 261	52 208	124	67 191	159
Eastern Asia..........	91 360	111 032	122	128 472	141
Southeastern Asia, major islands	84 223	96 472	115	120 392	143
China, Mainland	481 000	546 815	114	702 500	146
Total region	1 115 530	1 293 360	116	1 602 500	144
NEAR EAST					
Total region	90 750	104 280	115	132 370	146
AFRICA					
North Africa	17 617	21 276	121	26 814	152
West and central Africa	63 671	72 607	114	89 812	141
East and southern Africa	46 621	57 158	123	71 145	153
Total region	146 200	174 130	119	215 260	147
LATIN AMERICA					
Mexico and Central America	34 557	45 388	131	60 023	173
Northern and western countries of South America	29 018	37 133	128	47 424	163
Brazil	39 480	51 944	132	70 767	179
River Plate countries ..	16 894	20 923	124	25 319	150
Total region	125 075	161 726	129	211 026	169
EUROPE					
Western Europe	274 648	298 461	109	322 406	117
Eastern Europe (excl. U.S.S.R.)	92 909	87 315	94	95 795	103
U.S.S.R.	191 000	181 000	95	214 400	112
Total region	564 950	572 341	101	638 571	113

POPULATION 1938-60 (MIDYEAR ESTIMATES) AND INDEX NUMBERS OF POPULATION (1938 = 100), BY SUBREGIONS AND REGIONS (*Concluded*)

Subregions and regions	1938	1950		1960	
	Population in 1 000	*Population in 1 000*	*Index numbers*	*Population in 1 000*	*Index numbers*
NORTH AMERICA	141 470	166 041	117	198 563	140
OCEANIA	10 603	12 426	117	15 692	148
LOW-CALORIE COUNTRIES	1 460 661	1 712 573	117	2 135 837	146
HIGH-CALORIE COUNTRIES	733 917	771 73	105	878 145	120
WORLD	2 194 578	2 484 304	113	3 013 982	137

Appendix 2A

Index numbers of food production

(Prewar = 100)

Regions	Prewar	Average 1948/9-1952/3	1952/53	1953/54	1954/55	1955/56	1956/57	1957/58	1958/59	1959/60	1960/61	1961/62
Far East												
Excl. China, Mainland	100	106	112	121	122	126	131	129	135	142	146	147
Incl. China, Mainland	100	98	105	111	113	117	123	127	137	140	140	140
Near East	100	117	131	142	137	142	152	160	164	167	168	164
Africa	100	125	131	137	142	140	148	143	149	149	155	151
Latin America	100	126	134	137	144	145	155	159	167	167	168	171
Europe (incl. U.S.S.R.)	100	105	111	119	119	125	132	137	145	149	154	154
North America	100	139	149	148	146	153	157	152	163	165	167	165
Oceania	100	114	121	122	121	127	123	121	142	140	147	147
Less developed regions												
Excl. China, Mainland	100	115	122	129	131	134	141	142	148	152	155	155
Incl. China, Mainland	100	107	114	120	123	126	133	136	145	147	148	148
Developed regions ..	100	117	125	129	128	134	140	141	151	154	158	158
World excl. China, Mainland	100	116	124	129	130	134	141	141	150	153	157	157
Incl. China, Mainland	100	112	120	125	126	131	137	139	149	151	154	154

Percentage regional contribution to world food production

Regions	Prewar	1948/9-1952/3	1952/53	1953/54	1954/55	1955/56	1956/57	1957/58	1958/59	1959/60	1960/61	1961/62
Far East (incl. China, Mainland)	28	25	25	25	25	26	26	26	26	26	26	26
Near East	4	4	4	4	4	4	4	4	4	4	4	4
Africa	4	5	5	5	5	4	4	4	4	4	4	4
Latin America	7	7	7	7	8	7	8	8	8	8	7	7
Europe (incl. U.S.S.R.)	35	33	33	33	33	33	34	35	34	35	35	36
North America	20	24	24	23	23	23	22	21	22	21	22	21
Oceania	2	2	2	2	2	2	2	2	2	2	2	2
Less developed regions	43	41	41	41	42	41	42	42	42	42	41	41
Developed regions ..	57	59	59	59	58	59	58	58	58	58	59	59
World	100	100	100	100	100	100	100	100	100	100	100	100

Appendix 3A

INDEX NUMBERS OF PER CAPUT FOOD PRODUCTION

(PREWAR = 100)

Regions	Prewar	Average 1948/9-1952/3	1952/53	1953/54	1954/55	1955/56	1956/57	1957/58	1958/59	1959/60	1960/61	1961/62
FAR EAST												
Excl. China, Mainland	100	88	90	95	94	96	97	94	97	99	100	98
Incl. China, Mainland	100	83	85	88	88	90	92	92	98	97	95	93
NEAR EAST	100	98	105	112	105	106	111	114	115	114	112	107
AFRICA	100	101	102	104	105	101	105	100	102	100	101	96
LATIN AMERICA	100	94	95	95	98	96	100	100	103	100	98	98
EUROPE (incl. U.S.S.R.)	100	101	106	112	111	115	120	123	129	131	134	133
NORTH AMERICA	100	116	121	117	114	117	118	112	118	117	117	114
OCEANIA	100	94	96	94	91	93	89	85	98	94	97	94
LESS DEVELOPED REGIONS												
Excl. China, Mainland	100	93	96	100	99	99	102	101	103	103	103	101
Incl. China, Mainland	100	88	91	94	94	94	97	97	102	101	99	97
DEVELOPED REGIONS ..	100	110	114	116	115	119	122	121	128	129	130	129
WORLD excl. China, Mainland	100	100	103	106	105	107	110	108	113	113	114	111
Incl. China, Mainland	100	97	100	102	101	103	106	105	110	110	109	107

Appendix 3B

REGIONAL LEVELS OF PER CAPUT FOOD PRODUCTION AS PERCENTAGES OF THE WORLD PER CAPUT AVERAGE LEVELS

Regions	Prewar	1948/9-1952/3	1952/53	1953/54	1954/55	1955/56	1956/57	1957/58	1958/59	1959/60	1960/61	1961/62
Far East (incl. China, Mainland)	56	48	48	48	49	49	49	49	50	50	49	49
Near East	86	87	91	94	90	89	90	94	90	90	88	86
Africa	63	66	65	64	66	62	63	60	59	58	59	57
Latin America	119	116	114	111	115	111	113	114	111	109	107	109
Europe (incl. U.S.S.R.)	136	142	143	148	148	151	154	158	158	161	165	168
North America	302	362	364	346	340	342	336	321	324	322	322	320
Oceania	486	474	468	448	437	441	407	394	433	416	429	428
Less developed regions	64	58	58	59	59	59	59	59	59	59	58	58
Developed regions	174	194	198	198	197	200	200	200	202	203	207	208
World	100	100	100	100	100	100	100	100	100	100	100	100

PER CAPUT FOOD SUPPLIES AND NUTRITIVE V

Subregions and regions	Period	Cereals [1]	Starchy roots [2]	Sugar [3]	Pulses and nuts [4]	Vegetables and fruits [5]	Meat [6]	Eggs
		Kilograms per year						
FAR EAST (incl. China, Mainland)	Prewar	155	30	7	22	57	8	1
	Postwar	139	33	6	19	48	7	1
	Recent	146	61	9	18	53	9	1
South Asia..........	Prewar	139	8	13	22	51	3	–
	Postwar	121	7	12	21	30	2	–
	Recent	139	10	16	21	31	2	–
Southeastern Asia, mainland	Recent	149	17	7	24	53	10	2
Eastern Asia	Prewar	154	51	13	16	85	4	2
	Postwar	155	63	4	7	77	3	1
	Recent	156	60	12	15	94	6	4
Southeastern Asia, major islands	Prewar	129	111	7	9	63	7	1
	Postwar	122	100	4	10	50	6	1
	Recent	119	136	10	25	49	7	1
China, Mainland	Prewar	172	30	1	25	57	13	2
	Postwar	153	36	1	22	57	11	1
	Recent	153	89	3	15	61	14	1
NEAR EAST	Prewar	176	4	8	13	98	11	2
	Postwar	165	10	9	10	112	11	2
	Recent	163	16	14	17	145	13	2
AFRICA	Recent	121	173	11	14	78	15	1
North Africa	Recent	148	18	25	6	86	19	4
West and central Africa	Recent	93	320	4	14	94	6	1
East and southern Africa	Recent	149	36	15	15	58	25	2

s, BY SUBREGIONS AND REGIONS (RETAIL LEVEL)

Fish [8]	Milk [9]	Fats and oils [10]	Calories	Calories derived from cereals, starchy roots, and sugar	Calorie requirement	Calories derived from cereals, starchy roots, and sugar as percentage of total calories	Total calorie intake as percentage of calorie requirement	Total protein	Animal protein
......	......	......	...... Number per day	......	......	...... Percent	......	.. Grams per day	..
3	24	4	2 090	1 625	2 300	78	91	61	7
3	18	5	1 890	1 490	2 300	79	82	54	6
5	20	3	2 060	1 660	2 300	81	90	56	8
1	64	3	1 950	1 485	2 300	76	85	52	8
1	47	3	1 720	1 310	2 300	76	75	46	6
1	50	4	1 970	1 530	2 300	78	86	50	7
15	9	3	2 030	1 579	2 260	78	90	49	13
10	3	1	2 030	1 725	2 350	85	86	54	8
12	4	1	1 900	1 700	2 350	90	81	49	9
19	14	3	2 180	1 750	2 370	80	92	65	15
5	2	6	2 020	1 675	2 270	83	89	46	6
4	3	6	1 900	1 540	2 270	81	84	42	5
6	5	4	2 070	1 680	2 270	81	91	45	7
4	–	6	2 230	1 712	2 300	77	97	72	7
3	–	7	2 030	1 590	2 300	78	88	63	6
3	2	3	2 100	1 743	2 300	83	91	61	7
1	89	4	2 295	1 800	2 400	78	96	72	12
2	84	4	2 220	1 723	2 400	78	92	69	12
2	78	7	2 470	1 768	2 400	72	103	76	14
3	35	7	2 360	1 740	2 340	74	101	61	11
1	70	5	2 260	1 694	2 340	75	97	66	16
4	10	9	2 360	1 757	2 300	74	103	50	5
6	57	4	2 380	1 734	2 360	73	101	69	17

Subregions and regions	Period	Cereals [1]	Starchy roots [2]	Sugar [3]	Pulses and nuts [4]	Vegetables and fruit [5]	Meat [6]	Eggs [7]
		....	....	....	....	*Kilograms per year*	....	....
LATIN AMERICA	Prewar	90	75	25	15	90	48	4
	Postwar	101	84	30	14	100	35	3
	Recent	104	82	33	18	128	37	4
Mexico and Central America	Prewar	104	39	21	11	96	26	3
	Postwar	119	34	29	13	98	20	3
	Recent	114	32	32	19	110	23	5
Mexico	Prewar	109	5	18	9	67	25	3
	Postwar	129	7	30	12	60	20	4
	Recent	124	8	33	21	79	24	7
Central America	Prewar	93	43	23	11	195	34	3
	Postwar	135	11	27	11	116	18	2
	Recent	119	11	30	14	129	16	4
Caribbean	Prewar	93	151	33	16	157	29	5
	Postwar	84	115	29	17	168	21	3
	Recent	83	111	31	17	171	26	3
Northern and western countries of South America	Prewar	87	89	29	12	104	27	3
	Postwar	88	103	37	8	124	25	3
	Recent	84	97	36	10	151	28	3
Brazil	Prewar	78	91	25	23	88	50	3
	Postwar	86	112	25	24	82	27	2
	Recent	106	118	31	27	133	29	3
River Plate countries	Prewar	100	75	27	4	67	108	7
	Postwar	120	95	33	3	101	112	7
	Recent	112	80	32	3	116	105	7
EUROPE (incl. U.S.S.R.)	Prewar	151	132	18	9	99	34	6
	Postwar	142	155	19	8	105	29	5
	Recent	137	138	28	6	116	40	8

, BY SUBREGIONS AND REGIONS (RETAIL LEVEL) (*Continued*)

ish [8]	Milk [9]	Fats and oils [10]	Calories	Calories derived from cereals, starchy roots, and sugar	Calorie requirement	Calories derived from cereals, starchy roots, and sugar as percentage of total calories	Total calorie intake as percentage of calorie requirement	Total protein	Animal protein
....			...... *Number per day*		......	...... *Percent*		.. *Grams per day* ..	
2	90	6	2 160	1 347	2 420	63	89	64	28
3	68	8	2 315	1 530	2 410	66	96	62	22
3	82	9	2 510	1 580	2 410	63	104	67	24
2	77	6	1 950	1 341	2 430	69	80	54	18
2	66	7	2 230	1 570	2 420	71	92	57	15
2	85	9	2 370	1 540	2 420	65	98	63	19
1	86	5	1 800	1 265	2 450	70	74	53	18
2	67	7	2 220	1 610	2 450	73	90	58	16
2	94	9	2 440	1 590	2 450	85	100	68	20
1	18	6	1 965	1 214	2 340	62	84	51	16
1	61	4	2 190	1 630	2 370	75	92	58	13
1	69	5	2 130	1 510	2 370	71	90	58	14
4	68	7	2 475	1 639	2 410	66	102	60	21
2	67	8	2 280	1 450	2 390	64	95	54	16
3	78	11	2 410	1 450	2 390	60	101	57	19
2	71	4	1 970	1 355	2 510	69	78	55	18
3	66	5	2 160	1 500	2 480	70	87	53	18
5	76	7	2 190	1 440	2 480	66	88	56	20
1	81	5	2 190	1 301	2 310	59	95	64	28
2	34	7	2 180	1 420	2 310	65	95	55	15
2	60	8	2 650	1 700	2 310	64	115	67	19
2	161	9	2 740	1 457	2 560	53	107	95	59
2	162	15	3 150	1 700	2 560	56	123	106	63
2	141	16	3 040	1 640	2 560	54	119	96	55
6	131	13	2 870	1 930	2 590	67	111	85	28
6	143	11	2 760	1 883	2 590	68	107	82	29
7	180	16	3 040	1 903	2 590	63	117	88	36

Subregions and regions	Period	Cereal [1]	Starchy roots [2]	Sugar [3]	Pulses and nuts [4]	Vegetables and fruit [5]	Meat [6]	Egg
		Kilograms per year						
Western Europe	Prewar	130	107	23	9	112	44	[illegible]
	Postwar	124	118	23	8	125	33	[illegible]
	Recent	111	101	30	8	145	45	[illegible]
Eastern Europe and U.S.S.R.	Prewar	174	159	13	9	82	23	[illegible]
	Postwar	161	196	15	8	83	24	[illegible]
	Recent	166	179	26	3	82	35	[illegible]
NORTH AMERICA	Prewar	91	66	44	7	198	71	[illegible]
	Postwar	77	54	42	8	205	81	[illegible]
	Recent	67	49	41	7	188	91	[illegible]
OCEANIA	Prewar	98	49	52	2	139	118	[illegible]
	Postwar	96	50	52	5	157	109	[illegible]
	Recent	89	53	49	4	141	114	[illegible]
Low-calorie countries	Prewar	147	49	9	20	44	12	[illegible]
	Postwar	134	51	8	18	55	9	[illegible]
	Recent	141	71	11	18	67	11	[illegible]
High-calorie countries	Prewar	136	116	24	8	118	45	[illegible]
	Postwar	126	130	25	8	128	44	[illegible]
	Recent	119	114	32	6	133	55	[illegible]
WORLD	Prewar	144	70	14	17	79	23	[illegible]
	Postwar	132	76	14	15	78	20	[illegible]
	Recent	134	84	18	14	87	24	[illegible]

NOTE: Prewar and postwar totals for low-calorie countries and world include estimates for Africa.

[1] Cereals: In terms of flour and milled rice.
[2] Starchy roots: Includes sweet potatoes, cassava and other edible roots.
[3] Sugar: Includes raw sugar; excludes syrups and honey.
[4] Pulses and nuts: Includes cocoabeans.
[5] Vegetables and fruit: In terms of fresh equivalent.

sh [8]	Milk [9]	Fats and oils [10]	Calories	Calories derived from cereals, starchy roots, and sugar	Calorie requirement	Calories derived from cereals, starchy roots, and sugar as percentage of total calories	Total calorie intake as percentage of calorie requirement	Total protein	Animal protein
....			...... *Number per day*			...... *Percent*		.. *Grams per day* ..	
8	155	17	2 880	1 713	2 570	60	112	85	36
8	164	16	2 750	1 676	2 580	61	107	82	33
8	188	20	2 910	1 594	2 580	55	113	83	39
4	105	8	2 850	2 162	2 600	76	110	84	20
4	120	6	2 780	2 111	2 600	76	108	82	24
6	172	12	3 180	2 242	2 600	71	122	94	33
5	247	21	3 260	1 566	2 590	48	125	86	51
5	293	20	3 170	1 366	2 590	43	122	91	61
5	304	21	3 110	1 255	2 590	40	120	93	66
5	176	16	3 290	1 647	2 610	50	126	103	67
5	210	15	3 250	1 638	2 610	50	124	98	66
4	209	16	3 250	1 548	2 610	48	125	94	62
3	36	4	2 110	1 626	2 320	77	91	62	10
3	29	5	1 960	1 525	2 320	78	84	56	8
4	29	4	2 150	1 668	2 320	78	93	58	9
5	157	14	2 950	1 833	2 580	62	115	85	34
6	177	13	2 860	1 762	2 580	62	111	85	37
7	208	18	3 050	1 741	2 580	57	116	90	44
3	75	8	2 380	1 692	2 400	71	99	69	18
4	75	8	2 240	1 598	2 400	71	93	64	18
5	82	8	2 420	1 685	2 400	70	101	68	20

[6] Meat: Includes offal, poultry and game expressed in terms of carcass weight, excluding slaughter fats.
[7] Eggs: Fresh egg equivalent.
[8] Fish: Estimated edible weight.
[9] Milk: Excludes butter except for south Asia (India and Pakistan); includes milk products as fresh milk equivalent.
[10] Fats and oils: Pure fat content.

SHORT-TERM TARGETS FOR CALORIES, TOTAL PROTEINS, AND ANIMAL PROTEINS, BY SUBREGIONS AND REGIONS (PER CAPUT PER DAY)

Regions	Total calories (number)	Total proteins	Animal proteins	Animal proteins as percentage of total proteins	Percentage of total calories derived from proteins
		...... *Grams*			
FAR EAST					
South Asia	2 300	63	12.5	20	11
Southeastern Asia, mainland	2 300	57	15	26	10
Eastern Asia	2 350	72	20	28	12
Southeastern Asia, major islands	2 300	55	12	22	10
China, Mainland	2 350	72	11	15	12
REGION	2 300	68	12.5	18	12
NEAR EAST					
REGION	2 450	77	20	26	12
AFRICA					
North Africa	2 350	72	22	31	12
West and central Africa ..	2 400	63	13	21	10
East and southern Africa	2 450	73	21	29	12
REGION	2 400	69	18	26	12
LATIN AMERICA (excl. River Plate countries)					
Mexico and Central America	2 450	69	23	33	11
Northern and western countries of South America	2 500	72	25	35	12
Brazil	2 650	71	25	35	11
REGION	2 550	71	25	35	11
LOW-CALORIE COUNTRIES	2 350	69	15	22	12
WORLD	2 550	75	23	31	12

Appendix 6

PER CAPUT FOOD SUPPLIES NEEDED UNDER THE SHORT-TERM TARGET TOGETHER WITH (

Subregions and regions		Cereals	Starchy roots	Sugar	Pulses and nuts	Vegeta-bles and fruit	Meat	Eggs
FAR EAST		....				*Kilograms per year*		......
South Asia	Target ...	139	16	20	31	58	3	1
	% Change	99	167	127	149	184	160	300
Southeastern Asia, mainland	Target ...	149	16	8	40	93	11	3
	% Change	100	100	117	172	178	115	150
Eastern Asia	Target ...	151	60	18	18	104	8	5
	% Change	97	100	150	122	110	124	150
Southeastern Asia, major islands	Target ...	131	110	12	31	91	9	3
	% Change	110	81	114	128	185	139	200
China, Mainland	Target ...	150	87	7	24	94	23	3
	% Change	98	98	225	155	155	159	350
REGION	Target ...	144	58	13	27	82	13	2
	% Change	99	96	146	150	156	150	200
NEAR EAST								
REGION	Target ...	146	16	16	17	145	19	4
	% Change	90	100	119	100	100	151	200
AFRICA								
North Africa	Target ...	141	18	25	10	86	27	7
	% Change	95	100	100	159	100	140	167
West and central Africa	Target ...	120	160	4	19	94	15	2
	% Change	130	50	120	131	100	247	300
East and southern Africa	Target ...	146	27	15	15	73	27	4
	% Change	98	75	100	100	127	110	200
REGION	Target ...	133	88	11	16	85	22	4
	% Change	111	51	107	119	108	150	250

E AND PROTEIN LEVELS, BY SUBREGIONS AND REGIONS (RETAIL LEVEL)

Fish	Milk	Fats and oils	Calories	Percentage of calories derived from cereals, starchy roots, and sugar	Total proteins	Animal proteins	Over-all index of per caput food supply	Over-all index of per caput animal food supply
....	....	....	*Number per day*		.. *Grams per day* ..			
4	88	7	2 310	69	63	12.5	138	183
300	174	182						
18	12	5	2 300	69	57	15	121	122
120	148	175						
23	28	4	2 340	76	72	20	117	135
124	195	171						
12	11	5	2 310	74	55	12	130	162
178	231	150						
5	5	5	2 330	75	72	11	132	169
175	325	163						
8	36	6	2 310	73	68	12.5	131	163
175	181	178						
3	112	9	2 470	66	77	20	117	148
125	143	125						
2	89	5	2 330	70	72	22	117	142
250	129	100						
9	24	9	2 390	68	63	13	124	253
250	258	100						
7	73	5	2 440	68	73	21	114	124
133	128	125						
7	53	7	2 400	68	69	18	123	171
250	151	105						

Appendix 6

PER CAPUT FOOD SUPPLIES NEEDED UNDER THE SHORT-TERM TARGET TOGETHER WITH

Subregions and regions		Cereals	Starchy roots	Sugar	Pulses and nuts	Vegeta-bles and fruit	Meat	Egg
LATIN AMERICA (excl. River Plate countries)		....	....	....	....	*Kilograms per year*	....	.....
Mexico and Central America	Target ...	120	32	23	19	110	27	8
	% Change	106	100	72	100	100	119	150
Northern and western countries of South America	Target ...	120	59	27	14	151	31	4
	% Change	143	61	76	141	100	112	150
Brazil	Target ...	106	90	29	24	133	34	5
	% Change	100	76	94	88	100	118	156
REGION	Target ...	115	62	27	19	130	31	6
	% Change	112	75	83	100	100	118	145
LOW-CALORIE COUNTRIES	Target ...	141	59	14	25	91	16	3
	% Change	100	34	126	142	135	147	200
WORLD	Target ...	134	76	19	19	103	28	5
	% Change	100	90	108	137	118	115	136

E AND PROTEIN LEVELS, BY SUBREGIONS AND REGIONS (RETAIL LEVEL) (*Concluded*)

Fish	Milk	Fats and oils	Calories	Percentage of calorie derived from cereals, starchy roots, and sugar	Total proteins	Animal proteins	Over-all index of per caput food supply	Over-all index of per caput animal food supply
....	....	....	*Number per day*		.. *Grams per day* ..			
4	97	9	2 430	63	69	23	113	127
200	114	108						
9	89	8	2 500	65	72	25	116	124
192	117	122						
5	88	9	2 660	60	71	25	112	132
200	145	118						
6	91	9	2 550	62	71	25	114	128
200	124	114						
7	47	7	2 350	71	69	15	127	157
167	161	150						
7	95	10	2 560	66	75	23	111	116
136	116	123						

Appendix 7

POPULATION (1975 AND 2000) AND INDEX NUMBERS OF POPULATION (1958 = 100), BY SUBREGIONS AND REGIONS (MEDIUM ASSUMPTION)

Subregions and regions	1958 Population in millions	1975 Population in millions	1975 Index numbers	2000 Population in millions	2000 Index
FAR EAST					
South Asia	510	716	140	1 271	249
Southeastern Asia, mainland	64	92	143	162	254
Eastern Asia	125	161	129	231	185
Southeastern Asia, major islands	114	131	115	290	254
China, Mainland	670	949	142	1 685	251
REGION	1 531	2 150	140	3 753	245
NEAR EAST					
REGION	127	182	143	326	257
AFRICA					
North Africa	25	38	149	73	287
West and central Africa	85	111	131	178	210
East and southern Africa	68	91	134	153	225
REGION	205	273	133	458	223
LATIN AMERICA (excl. River Plate countries)					
Mexico and Central America	57	89	157	181	319
Northern and western countries of South America	45	68	151	135	301
Brazil	66	104	159	219	334
REGION	175	272	155	552	315
RIVER PLATE COUNTRIES	25	32	128	43	172
EUROPE					
REGION	624	757	121	954	153
NORTH AMERICA					
REGION	192	250	130	325	169
OCEANIA					
REGION	15	22	147	30	200
LOW-CALORIE COUNTRIES	2 038	2 877	141	5 089	250
HIGH-CALORIE COUNTRIES	856	1 061	124	1 352	158
WORLD	2 894	3 938	136	6 441	223

Appendix 8

INDEX NUMBERS OF NEEDS IN TOTAL FOOD SUPPLIES BY 1975 UNDER THE SHORT-TERM TARGET, BY SUBREGIONS AND REGIONS

Subregions and regions	Cereals	Starchy roots	Sugar	Pulses and nuts	Vegetables and fruit	Meat	Eggs	Fish	Milk	Fats and oils	Total food	Animal food
FAR EAST												
South Asia	139	234	178	209	258	224	420	420	244	255	193	256
Southeastern Asia, mainland	143	143	167	246	255	164	215	172	212	250	173	174
Eastern Asia	125	129	194	157	142	160	194	160	252	221	151	174
Southeastern Asia, major islands	127	93	131	147	213	160	230	205	266	173	150	186
China, Mainland	139	139	320	220	220	226	497	249	462	231	187	240
REGION	139	134	204	210	218	210	280	245	253	249	183	228
NEAR EAST												
REGION	129	143	170	143	143	216	286	179	204	179	167	212
AFRICA												
North Africa........	142	149	149	237	149	209	249	373	192	149	174	212
West and central Africa	170	66	157	172	131	324	393	328	338	131	162	331
East and southern Africa	131	101	134	134	170	147	268	178	172	168	153	166
REGION	148	68	142	158	144	200	333	333	201	140	164	227

Appendix 8

INDEX NUMBERS OF NEEDS IN TOTAL FOOD SUPPLIES BY 1975 UNDER THE SHORT-THERM TARGET, BY SUBREGIONS AND REGIONS

Subregions and regions	Cereals	Starchy roots	Sugar	Pulses and nuts	Vegeta-bles and fruits	Meat	Eggs	Fish	Milk	Fats and oils	Total food	Animal food
LATIN AMERICA (excl. River Plate countries)												
Mexico and Central America	166	157	113	157	157	187	236	314	179	170	177	199
Northern and western countries of South America	216	92	115	213	151	169	227	290	177	184	175	187
Brazil	159	121	149	140	159	188	248	318	231	188	178	210
REGION	174	116	129	155	155	183	225	310	192	177	177	198
LOW-CALORIE COUNTRIES	141	118	178	200	190	207	282	235	227	212	179	221
WORLD	136	122	147	186	160	156	185	185	158	167	151	158

World Food Supply

An Arno Press Collection

Agricultural Production Team. **Report on India's Food Crisis & Steps to Meet It.** 1959

Agricultural Tribunal of Investigation. **Final Report.** Presented to Parliament by Command of His Majesty. 1924

Bennett, M. K. **The World's Food:** A Study of the Interrelations of World Populations, National Diets and Food Potentials. 1954

Bhattacharjee, J. P., editor. **Studies in Indian Agricultural Economics.** 1958

Brown, Lester R. **Increasing World Food Output:** Problems and Prospects. 1965

Brown, Lester R. **Man, Land & Food:** Looking Ahead at World Food Needs. 1963

Christensen, Raymond P. **Efficient Use of Food Resources in the United States.** Revised Edition. 1948

Crookes, William. **The Wheat Problem.** Revised Edition. 1900

Developments in American Farming. 1976

Dodd, George. **The Food of London.** 1856

Economics and Sociology Department, Iowa State College. **Wartime Farm and Food Policy,** Pamphlets 1-11. 1943/44/45

Edwards, Everett E., compiler and editor. **Jefferson and Agriculture:** A Sourcebook. 1943

Famine in India. 1976

Gray, L. C., et al. **Farm Ownership and Tenancy.** 1924

Hardin, Charles M. **Freedom in Agricultural Education.** 1955

High-Yielding Varieties of Grain. 1976

[India] Famine Inquiry Commission. **Report on Bengal.** 1945

Johnson, D. Gale. **Forward Prices for Agriculture.** With a New Introduction. 1947

King, Clyde L., editor. **The World's Food.** 1917

Marston, R[obert] B[right]. **War, Famine and our Food Supply.** 1897

Mosher, Arthur T. **Technical Co-operation in Latin-American Agriculture.** 1957

The Organization of Trade in Food Products: Three Early Food and Agriculture Organization Proposals. 1976

Projections of United States Agricultural Production and Demand. 1976

Rastyannikov, V. G. **Food For Developing Countries in Asia and North Africa:** A Socio-Economic Approach. Translated by George S. Watts. 1969

Reid, Margaret G. **Food For People.** 1943

Schultz, Theodore W., editor. **Food For the World.** 1945

Schultz, Theodore W. **Transforming Traditional Agriculture.** 1964

Three World Surveys by the Food and Agriculture Organization of the United Nations. 1976

U. S. Department of Agriculture, Agricultural Adjustment Administration. **Agricultural Adjustment:** A Report of Administration of the Agricultural Adjustment Act, May 1933 To February 1934. 1934

U. S. Department of Agriculture. **Yearbook of Agriculture, 1939:** Food and Life; Part 1: Human Nutrition. 1939

U. S. Department of Agriculture. **Yearbook of Agriculture, 1940:** Farmers in A Changing World. 1940

[U. S.] House of Representatives, Committee on Agriculture. **Oleomargarine.** 1949

[U. S.] National Resources Board. **Report of the Land Planning Committee. Part II.** 1934